APPLICATIONS OF BIOTECHNOLOGY TO TRADITIONAL FERMENTED FOODS

Report of an Ad Hoc Panel of the
Board on Science and Technology
for International Development

Office of International Affairs
National Research Council

NATIONAL ACADEMY PRESS
Washington, D.C. 1992

NOTICE: The project that is the subject of this report was approved by the Governing Board of the National Research Council, whose members are drawn from the councils of the National Academy of Sciences, the National Academy of Engineering, and the Institute of Medicine. The members of the committee responsible for the report were chosen for their special competence and with regard for appropriate balance.

This report has been reviewed by a group other than the authors according to procedures approved by a Report Review Committee consisting of members of the National Academy of Sciences, the National Academy of Engineering, and the Institute of Medicine.

The National Academy of Sciences is a private, nonprofit, self-perpetuating society of distinguished scholars engaged in scientific and engineering research, dedicated to the furtherance of science and technology and to their use for the general welfare. Upon the authority of the charter granted to it by the Congress in 1863, the Academy has a mandate that requires it to advise the federal government on scientific and technical matters. Dr. Frank Press is president of the National Academy of Sciences.

The National Academy of Engineering was established in 1964, under the charter of the National Academy of Sciences, as a parallel organization of outstanding engineers. It is autonomous in its administration and in the selection of its members, sharing with the National Academy of Sciences the responsibility for advising the federal government. The National Academy of Engineering also sponsors engineering programs aimed at meeting national needs, encourages education and research, and recognizes the superior achievements of engineers. Dr. Robert M. White is president of the National Academy of Engineering.

The Institute of Medicine was established in 1970 by the National Academy of Sciences to secure the services of eminent members of appropriate professions in the examination of policy matters pertaining to the health of the public. The Institute acts under the responsibility given to the National Academy of Sciences by its congressional charter to be an adviser to the federal government and, upon its own initiative, to identify issues of medical care, research, and education. Dr. Stuart Bonderant is acting president of the Institute of Medicine.

The National Research Council was organized by the National Academy of Sciences in 1916 to associate the broad community of science and technology with the Academy's purposes of furthering knowledge and advising the federal government. Functioning in accordance with general policies determined by the Academy, the Council has become the principal operating agency of both the National Academy of Sciences and the National Academy of Engineering in providing services to the government, the public, and the scientific and engineering communities. The Council is administered jointly by both Academies and the Institute of Medicine. Dr. Frank Press and Dr. Robert M. White are chairman and vice chairman, respectively, of the National Research Council.

The Board on Science and Technology for International Development (BOSTID) of the Office of International Affairs addresses a range of issues arising from the ways in which science and technology in developing countries can stimulate and complement the complex processes of social and economic development. It oversees a broad program of bilateral workshops with scientific organizations in developing countries and conducts special studies. BOSTID's Advisory Committee on Technology Innovation publishes topical reviews of technical processes and biological resources of potential importance to developing countries.

This report has been prepared by an ad hoc advisory panel of the Advisory Committee on Technology Innovation, Board on Science and Technology for International Development, Office of International Affairs, National Research Council. Staff support was funded by the Office of the Science Advisor, Agency for International Development, under Grant No. DAN-5538-G-00-1023-00, Amendments 27 and 29.

Library of Congress Catalog Card Number: 91-68331
ISBN 0-309-04685-8
S526
Printed in the United States of America

COVER DESIGN by DAVID BENNETT

Panel on the Applications of Biotechnology to Traditional Fermented Foods

Elmer L. Gaden, Jr. (*Chairman*), Department of Chemical Engineering, University of Virginia, Charlottesville, Virginia

Mpoko Bokanga, International Institute of Tropical Agriculture, Ibadan, Nigeria.

Susan Harlander, Department of Food Science and Nutrition, University of Minnesota, St. Paul, Minnesota

Clifford W. Hesseltine, Northern Regional Research Center, U.S. Department of Agriculture, Peoria, Illinois

Keith H. Steinkraus, Institute of Food Science, Cornell University, Ithaca, New York

Advisory Group

K. E. Aidoo, University of Strathclyde, Glasgow, United Kingdom

Samuel Angel, Agricultural Research Organization, Bet Dagan, Israel

Mogessie Ashenafi, Awassa College of Agriculture, Awassa, Ethiopia

E. V. Carpio, Institute of Food Science and Technology, University of the Philippines at Los Baños, Philippines

Hamid A. Dirar, Faculty of Agriculture, University of Khartoum, Sudan

Sara Feresu, University of Zimbabwe, Harare, Zimbabwe

Abed Hamama, Institut Agronomique et Veterinaire, Hassan II, Rabat-Instituts, Morocco

David B. Harper, Queen's University of Belfast, Belfast, Northern Ireland, United Kingdom

Hiroshi Motai, Research Division, Kikkoman Corporation, Chiba, Japan

Felixtina E. Jonsyn, Njala University College, Freetown, Sierra Leone

J. Maud Kordylas, Arkloyd's Food Laboratory, Douala, Cameroon

M. Kroger, The Pennsylvania State University, University Park, Pennsylvania

J. A. Kurman, Agricultural Institute, Grangeneuve, Switzerland

L. B. Mabesa, Institute of Food Science and Technology, University of the Philippines at Los Baños, Philippines

Reynaldo Mabesa, Institute of Food Science and Technology, University of the Philippines at Los Baños, Philippines

Nguyen Hoai Huong, Institute for Experimental Biology, Ho Chi Minh City, Vietnam

NGUYEN NGOC THAO, Institute for Experimental Biology, Ho Chi Minh City, Vietnam

M. J. R. NOUT, Food Science Department, Agricultural University, Wageningen, The Netherlands

NDUKA OKAFOR, University of Nigeria, Nsukka, Nigeria

MINERVA SD. OLYMPIA, Institute of Fish Processing Technology, College of Fisheries, University of the Philippines in Visayas, Iloilo, Philippines

O. B. OYEWOLE, University of Agriculture, Abeokuta, Nigeria

OCTAVIO PAREDES-LOPEZ, CIEA-Instituto Politecnico Nacional, Irapuato, Gto., Mexico

J. L. RASIC, Food Research Institute, Novi Sad, Yugoslavia

S. SALMINEN, Dairies Cooperative Association, Helsinki, Finland

TAKASHI HAMADA, Research Division, Kikkoman Corporation, Chiba, Japan

PAIROTE WIRIYACHAREE, Chiang Mai University, Chiang Mai, Thailand

MARGY J. WOODBURN, Oregon State University, Corvallis, Oregon

YAICHI FUKUSHIMA, Research Division, Kikkoman Corporation, Chiba, Japan

LESLIE FOOK-MIN YONG, Aroma Biotech Pte. Ltd., Singapore

National Research Council Staff

GRIFFIN SHAY, *Senior Program Officer and Staff Study Director*

F. R. RUSKIN, *Editor*

CONSTANCE REGES, *Administrative Secretary*

MICHAEL MCD. DOW, *Acting Director, Board on Science and Technology for International Development*

Contents

Preface

The purpose of this report is to create greater awareness of the opportunities to reduce hunger and improve nutrition in developing countries through the application of biotechnology to widely practiced methods of food preparation and preservation. The report discusses opportunities for the application of biotechnology to traditional fermented foods. Scientists from developed and developing countries describe their research in this field and provide their recommendations on priorities for future research.

Preparation of this report was coordinated by the Board on Science and Technology for International Development in response to a request from the U.S. Agency for International Development.

I. RESEARCH PRIORITIES

Research Priorities in Traditional Fermented Foods

The Advisory Panel

Biotechnology has been described as the application of scientific and engineering principles to the processing of materials for the provision of goods and services through the use of biological systems and agents. In a very real sense, biotechnology originated with traditional food fermentations in developing countries. Over the generations, this pioneering practice has been expanded and improved so that microorganisms and other biological agents have found use in many other areas. Recent developments in genetics, enzymology, recombinant technology, and fermentation technology have led to advances in biotechnology far beyond the original traditional scope.

In many developing countries, village-art methods and age-old techniques are still used for food processing. Developing countries appear to be neglecting the advances in biotechnology. But they cannot continue to depend on historic methods for food processing. Increasing populations, drought and other natural disasters, and inadequate food production dictate that better options for food processing be adopted. Biotechnology offers this opportunity.

Current food biotechnological research in developing countries seems largely limited to the identification of microorganisms for starter culture development. There is little research involving gene manipulation and there are few centers of operational biotechnological research. The reasons for this are obvious. Biotechnological research is capital intensive, usually in scarce foreign exchange. Also, biotechnology requires the use of sophisticated equipment and reagents backed with a consistent energy and water supply, which are often not available in developing countries. A crucial part or essential chemical—which should be no more than a telephone call away, and can be obtained, at most, overnight in industrialized countries—cannot be obtained in months or even years. Or, just when all the necessary personnel and materials are available, the electricity is cut off.

To meet the current and future challenges in developing countries, it is important that these countries develop the capabilities to benefit from biotechnological developments. Developing countries will need to acquire expertise in biotechnology through education and training. The infrastructure and equipment required for biotechnological research will need to be established. Scientists of the developing world will need to collaborate with laboratories in advanced countries in order to benefit from their knowledge and to obtain infrastructural support and funding. It is through these strategies that the earliest application of biotechnology can be enhanced through help from its heirs.

PRIORITIES

The recommended research priorities encompass four broad categories: (1) improving understanding of the fermentation processes; (2) refining of the processes; (3) increasing the utilization of the processes; and (4) developing local capabilities. In this research, special emphasis should be given to fermented products that serve as major sources of nourishment for large populations (cassava, for example), processes that reduce food loss, foods that may alleviate starvation in famine or drought, and foods for weaning and young children.

IMPROVING THE KNOWLEDGE BASE

For fermented products like cheese, bread, beer, and wine, scientific and technological knowledge of the processes is well developed. However, for traditional fermented products, this knowledge is poor. Many indigenous fermented foods are produced by spontaneous or natural fermentation, but specific microorganisms predominate. Isolation and characterization of predominant organisms is essential.

Information should be collected on all traditional fermented foods and it must be thorough. No food should be excluded because it is not important or well known. A thorough microbiological, nutritional, and technical investigation should be carried out on each of the processes. The various microorganisms involved in each fermentation should be isolated, characterized, studied, and preserved. The biotechnological worth of each organism should be determined. Isolation should not be confined to the dominant organisms because other microbes found in lower numbers might have an important function in the process. The role of each organism should be identified.

Much basic research is needed to determine the scientific and technological factors in the preparation of these traditional products. Since the qualities of fermented foods are largely controlled by the participating microorganisms, understanding their role is vital.

IMPROVING THE TECHNOLOGY

In food fermentations, raw materials are converted to products through the use of biocatalysts. Each member of this equation is important. For widely used plant substrates, for example, breeding to reduce toxic or antinutritional components, or to increase protein or vitamin content, would be useful. Alternatively or additionally, it would be valuable to identify microorganisms that can synthesize important ingredients (e.g., essential amino acids, vitamins) for populations where malnutrition is a problem. Some additional desirable traits for these microorganisms are: an ability to produce flavor components that which favor consumption of these foods in traditional and new markets; the capability to break down antinutritional factors (i.e., phytic acid) present in some substrates; the production of enzymes to utilize recalcitrant wastes as substrates; the inability to synthesize toxins and other undesirable secondary products; and thermotolerance and osmotolerance, which are important characteristics in solid substrate fermentation processes.

For lactic acid bacteria used in food fermentations, physiological characteristics of acid stability, bile stability, adherence to human intestinal cells, colonization of the human intestinal tract, and antagonism to pathogenic bacteria and cariogenic bacteria (oral health) are all desirable.

The safety and shelf life of fermented products may also be improved through the development of organisms that produce alcohols, antibiotics, or other substances that can inhibit the growth of undesirable organisms.

The art of traditional processes needs to be transformed into a technology to incorporate objective methods of process control and optimization, and to standardize quality of the end products without losing their desirable attributes. Fermentations can only be optimized when conditions like time, temperature, pH, substrate pretreatment, inoculum-substrate ratio, and so forth, are controlled. Because of the surface:volume relationships, the scale-up of solid state fermentations is particularly difficult. These solid state reactions can be valuable in reducing raw material losses.

The equipment needed for the improvement of some traditional processes can be a challenge in itself. Fermentations carried out in vessels with unusual surface characteristics such as charred wood, semi-porous clay, gourds, or the like, are difficult to replicate.

Research is also needed on the implementation of continuous fermentations using bioreactors with immobilized enzymes and cells. Research on the development of bioreactors with improved performance is required.

IMPROVING UTILIZATION

The introduction of new processes or products should take into account the sensory requirements of target social groups. Thus, the elucidation of the microbial origin of flavors in fermented foods and the relationship between microflora and the organoleptic properties of the product are imperative. Flavor and color must be generated to meet local population preferences.

The use of alternative plant materials such as triticale, oca, amaranth, and achira, which have been successfully grown in some developing countries, should be examined as substrates for fermentations. *Puto* is a fermented rice cake in the Philippines. In a taste test, *puto* in which cassava was substituted for half of the rice was preferred over pure rice *puto*. Acha (*Digitaria exilis*), a West African cereal crop also known as "fonio," and ensete (*Ensete ventricosum*) are being tested as alternative substrates for food fermentations. A major drawback of ensat is its low protein content (1.5 percent) compared with other cereals; a plus is that it contains twice as much methionine as maize and wheat. Acha is being examined for the production of traditional porridge, beer, pasta, and even bread. Studies of these less-known fermented products could lead to processes with minimum production cost and maximum substrate utilization, resulting in products with improved nutritional value, extended shelf life, improved quality, and a better spectrum of essential nutrients. Inclusion of soy or other vegetable proteins could also enhance the nutritive value of many low protein foods.

The ability to use alternative substrates could also reduce problems of sporadic nonavailability of traditional starting materials. Acceptability of new products or improvement of traditional ones could be improved through the distribution of starter cultures. Some cultures are difficult to maintain in dehydrated form, and this is an important area for research. Acceptability of fermented products based on alternative raw materials may hinge on using familiar processing steps such as roasting or germination.

Research on fermentations that use wastes as raw materials has several possible benefits. The use of agroindustrial residues and other wastes to produce fermented foods and feeds can optimize indigenous resources, increase the availability of nutritious products, and reduce pollution problems.

Research is also needed on improving the economics of fermentation processes. Reducing the time necessary to pretreat raw materials or the processing time can be valuable. It would be helpful, for example, to reduce the boiling time (6 to 8 hours) of sesame seed before fermentation. Reducing fermentation time can optimize equipment use.

DEVELOPING LOCAL CAPABILITIES

Biotechnology is possible only within an infrastructure of supply companies that can provide specialized equipment and reagents. In addition, there must be a constant source of electricity for continuing experiments, and often for the air conditioning necessary for the growth of specific organisms. Developing local or regional production of commonly used enzymes would help.

Training in basic microbiology, biochemical engineering, and the new techniques of molecular biology for personnel of less developed countries is one of the key components in improving traditional fermentation processes. In addition, developing country scientists would also benefit from opportunities for regional and international collaboration. This kind of information sharing could be facilitated through periodic seminars and workshops, through joint research programs, and through the establishment of computer networks. Each of these interactions could include scientists from industrialized countries. Centers of excellence, specializing in regionally important areas, could be established for the mutual benefit of cooperating institutions.

For large-scale fermentations, developing countries should give higher priority to industrializing appropriate indigenous processes, rather than importing the technology of the industrialized world. This imported technology often relies on imported crops or crops not well suited to the climate or soils of the country.

In modernizing the production of traditional fermented foods at the village level, appropriate and affordable technology should be emphasized. Process changes should take into account the role of the poor who originated and preserved the processes and how they will benefit from the modifications.

II. OVERVIEW

1

Upgrading Traditional Biotechnological Processes

M. J. R. Nout

TRADITIONAL FOOD FERMENTATION

The general aims of food technology are to exploit natural food resources as efficiently and profitably as possible. Adequate and economically sound processing, prolongation of shelf life by preservation and optimization of storage and handling, improvement of safety and nutritive value, adequate and appropriate packaging, and maximum consumer appeal are key prerequisites to achieving these aims.

Fermentation is one of the oldest methods of food processing. The history of fermented foods has early records in Southeast Asia, where China is regarded as the cradle of mold-fermented foods, and in Africa where the Egyptians developed the concept of the combined brewery-bakery. The early Egyptian beers were probably quite similar to some of the traditional opaque sorghum, maize, or millet beers found in various African countries today (1).

In technologically developed regions, the crafts of baking, brewing, wine making, and dairying have evolved into the large-scale industrial production of fermented consumer goods, including cheeses, cultured milks, pickles, wines, beers, spirits, fermented meat products, and soy sauces.

The introduction of such foreign "high-tech" fermented products to tropical countries by early travelers, clergymen, and colonists was followed by an accelerated demand during the early postindependence period. Their high price ensured status, and their refined quality guaranteed continued and increasing consumption.

In contrast, many of the traditional indigenous foods lack this image; some may even be regarded as backward or poor people's food. Factors contributing to such lack of appeal include inadequate grading and cleaning of raw materials, crude handling and processing techniques, and insufficient product protection due to lack of packaging. Such unhygienic practices are easily translated into a fear of food-borne diseases. From a nutritionist's point of view, many traditional starchy staples are deficient in energy, protein, and vitamins. Variable sensory characteristics (quality) and lack of durability (shelf life) reduce convenience to the consumer: time needs to be spent selecting products of adequate quality, whereas perishable products require frequent purchasing and result in increased wastage. In addition, ungraded heterogenous products, inconvenient unpacked bulk foods, or unattractive presentation inhibit consumers to develop regular purchasing attitudes.

The contrast outlined here serves as a general guideline to the major targets for upgrading the present status of traditional indigenous fermented foods. The latter are part of the regional cultural heritage; they are well known and accepted by consumers and consequently provide an appropriate basis for development of a local food industry, which not only preserves the agricultural produce but also stimulates and supports agroindustrial development.

DECENTRALIZED SMALL-SCALE PROCESSES

In most African countries, 70 percent or more of the population lives in rural areas. However, if the present trend in urbanization continues (urban growth rates of 5 to 10 percent annually), 50 percent of the African population will be living in cities by the year 2000. Governments become increasingly aware that rural industrialization is a worthwhile investment because it creates job opportunities, improves agricultural productivity, and helps to check urbanization. But even at the present urbanization rate, a rapidly increasing low-income population will be located in urban areas. The resultant uncoupling in place and time of primary production and food consumption necessitates the manufacture of wholesome, low-cost, nutritious products that can withstand low-hygiene handling.

Agro-allied industries are closely linked to regions of primary production, and it is particularly in the field of food processing, with low-cost perishable raw materials, that establishment of a rural network of small-scale processing facilities is most appropriate. Home- or village-scale enterprises require only modest capital investment, which should be made available on a "soft loan" basis. Against this background, some basic process improvements that increase the appeal of traditional fermented foods and that can be carried out by simple means will be outlined (2).

BASIC PROCESSING OPERATIONS

In food manufacturing several operations are required to prepare raw materials, handle and process them into products, and finally prepare the finished product for distribution and sale by preservation and/or packaging. One might think of sorting, grading, cleaning, disinfection, grinding, or packaging. The establishment and success of some indigenous enterprises in Nigeria and Kenya show that the appeal and marketability of such products as beans, peas, *gari*, and spices, formerly sold in bulk, increase significantly when they have "only" been sorted, cleaned, graded, sometimes ground, labeled, and packaged in simple polythene bags.

NUTRITIVE VALUE

The nutritive value of traditional fermented foods needs improvement. The energy density of starch-based porridges is inadequate, particularly when used for weaning purposes. Root crop- or cereal-derived products have rather low protein contents, and the quality of their protein is limited by the amount of lysine present. Various antinutritional factors, including polyphenols, phytic acid, trypsin inhibitors, and lectins, are present in legumes and cereals.

Composite products (legume additions to starchy staples) offer an opportunity to improve protein quantity and quality. Combinations of simple unit operations, including roasting, germination, and fermentation, afford increased energy density in porridges and reduce antinutritional factors considerably (3).

STABILIZATION OF NATURAL FERMENTATIONS BY INOCULUM ENRICHMENT

Most traditional fermented products result from natural fermentations carried out under nonsterile conditions. The environment resulting from the chemical composition of the raw materials, fermentation temperature, absence or presence of oxygen, and additives such as salt and spices causes a gradual selection of microorganisms responsible for the desired product characteristics.

The main advantage of natural fermentation processes is that they are fitting to the rural situation, since they were in fact created by it. Also, the consumer safety of several African fermented foods is improved by lactic acid fermentation, which creates an environment that is unfavorable to pathogenic Enterobacteriaceae and Bacillaceae.

In addition, the variety of microorganisms present in a fermented food can create rich and full flavors that are hard to imitate when using pure starter cultures under aseptic conditions.

However, natural fermentation processes tend to be difficult to control if carried out at a larger scale; moreover, the presence of a significant accompanying microflora can accelerate spoilage once the fermentation is completed. Particularly with increased holding periods between product fermentation and consumption when catering for urban markets, uncontrolled fermentations under variable conditions will cause unacceptable wastage by premature spoilage.

Techniques to stabilize fermentations operating under nonsterile conditions would therefore be appropriate in the control of natural fermentations. For this purpose the use of pure culture starters, obtained either by laboratory selection procedures or genetic engineering, offers no realistic solutions because they are expensive and require sterile processing conditions. A more feasible approach is to exploit the ecological principle of inoculum enrichment by natural selection. This can be achieved by the sourdough process, in which some portion of one batch of fermented dough is used to inoculate another batch. This practice is also referred to as "back-slopping" or inoculum enrichment. The resulting starters are active and should not be stored but used in a continuous manner.

Sourdoughs from commercial sources, having been maintained by daily or weekly transfers during 2 or more years, contain only two or three microbial species, although they are exposed to a wide variety of potential competitors and spoilage-causing microorganisms each time the sourdough is mixed with fresh flour for a transfer. It can take as long as 10 weeks of regular transfers before a sourdough population becomes stabilized. Such populations could contain a yeast, *Saccharomyces exiguous*, and one or two *Lactobacillus* species, namely *Lb. brevis* var. linderi II and *Lb. sanfrancisco*. Although the mechanism of the stable coexistence of sourdough populations is not yet fully understood, lack of competition for the same substrate might play an important role. Other factors besides substrate competition, such as antimicrobial substances produced by lactic acid bacteria, might play an important role in the stability of such stable populations, obtained by "back-slopping" (4).

Similar experiments in the field of *tempe* manufacture showed that the first stage of the *tempe* process—soaking of soybeans—can be rendered more predictable in terms of acidification of the beans, by simple inoculum enrichment. Depending on soaking temperatures, stable soaking water populations were obtained after 30 to 60 daily transfers, containing *Leuconostoc* spp. at 14° and 19°C, yeasts and *Lactobacillus* spp. at 25°C, *Lactobacillus* spp. at 30°C, or *Pediococcus* and *Streptococcus* spp. at 37° and 45°C. *Tempe* made with well-acidified beans contained fewer undesirable microorganisms and was more attractive (5).

Based on the same principle of inoculum enrichment, the intrinsic microbiological safety of composite meals of cereals and legumes can be improved significantly by lactic fermentation (6). This offers interesting possibilities in the manufacture of food for vulnerable consumer groups, such as infants, malnourished patients, and the elderly (7).

Although development of such gradually evolved and stable fermentation starters will be an attractive proposition for use in small-scale fermentations under nonsterile conditions, they will not be the most appropriate in all cases. This is exemplified by the sauerkraut (lactic acid fermented cabbage) fermentation, during which flavor development is determined by a succession of *Leuconostoc* and *Lactobacillus* species occurring during the course of the fermentation. Practical experience in the sauerkraut industry in the Netherlands has shown that carryover of previous sauerkraut into a fresh batch of cabbage will cause a rapid domination of homofermentative *Lactobacillus* spp., which should normally only dominate during the final stage of fermentation. The result is an excessively sour-tasting product that lacks the flavor otherwise produced by the heterofermentative *Leuconostoc* and *Lactobacillus* spp.

In the exercise of upgrading traditional food fermentation techniques, it would therefore be worthwhile to investigate the effect of inoculum enrichment on product characteristics and consumer acceptance.

MULTISTRAIN DEHYDRATED STARTER

A different tool to stabilize fermentations under nonsterile conditions is the use of multistrain dehydrated starters, which can be stored at ambient temperatures, enabling more flexibility. Such homemade starters are widely used in several Asian food fermentations. Examples are the manufacture of *tempe* (mainly from soybeans) and *tapé* (from glutinous rice or cassava). Indonesian traditional *tempe* starters (*usar*) are essentially molded hibiscus leaves that carry a multitude of molds, dominated by *Rhizopus* spp., including the *Rh. oryzae* and *Rh. microsporus* varieties. Instead of using *usar*, Indonesian *tempe* production is increasingly carried out with factory-prepared "pure" starters consisting of granulated cassava or soybean fiber carrying a mixed population of *Rhizopus* species (5). These starters are more homogenous and their dosage is convenient, but because they are manufactured under nonsterile conditions, some are heavily contaminated with spoilage-causing bacteria and yeasts. This requires quality monitoring of the inoculum and of the fermentation process in which it is used.

Other examples of durable home-prepared starter materials used in Asian food fermentations are Indonesian *ragi* and Vietnamese *men* tablets (8). Depending on their specific purpose, these dehydrated tablets, prepared from fermented rice flour, contain mixed populations of yeasts, molds, and bacteria. *Ragi* tablets can be stored up to 6 months and constitute a convenient starter material for application in home and small-scale industrial fermentations of rice or cassava, for example.

Especially in the fermentation of neutral pH, protein-rich substrates, such as legumes, one should be extremely careful with the use of substandard inoculum. If the process lacks factors that control microbial development, pathogens may survive or produce toxins in such products. *Tempe* manufacture is a good example of a process with intrinsic safety. The preliminary soaking of the beans results in an acidification that inhibits the multiplication of bacterial contaminants during the mold fermentation stage. Also, antimicrobial substances of *Rhizopus oligosporus* would play a protective role against outgrowth of several genera of microorganisms. Moreover, near-anaerobic conditions and microbial competition during the fermentation stage, and the usual cooking or frying of *tempe* prior to consumption, strongly reduce the chances of food-borne illness (5).

Nevertheless, the introduction of fermentation processes in regions where they are not traditionally mastered requires adequate guidance, supervised processing, and monitoring of product safety.

ENZYME PRODUCTION BY *KOJI* TECHNIQUE

Not only microorganisms but also enzymes play an important role in the manufacture of traditional fermentation processes. In cassava processing the naturally occurring enzyme linamarase is able to degrade potentially toxic cyanogenic glycosides (e.g., linamarin). This enzymatic detoxification has always been an integral part of traditional cassava fermentations, such as in *gari* and *lafun*. Under certain conditions the detoxification of linamarin is accelerated by linamarase addition (9). It is conceivable that there will be commercial applications for the enzymatic process of linamarin decomposition, which could be used to detoxify cassava without having to ferment it; the result would be a neutral and bland-flavored product.

Enzyme sources for African traditional beer brewing are mostly germinated sorghum and millet varieties, whereas sorghum and millet malts possess adequate diastatic power with α-amylase, resulting in poor conversion of dextrins into maltose (10). The availability of cheap technical-grade ß-amylase preparations could lead to the development of novel brewing processes utilizing home-grown starch sources instead of imported barley malt.

In East Asia, *koji* is used as a source of enzymes in the manufacture of soy sauce and rice wine. *Koji* is obtained by solid-substrate fermentation of cereals or soybeans with fungi (e.g., *Aspergillus oryzae* and *Asp. soyae*). Depending on the particular substrate to be degraded, selected strains of molds are used, often as mixed cultures. Their enzymes include amylases, proteases, and cellulolytic enzymes. During fermentation the enzymes are accumulated into the *koji*. The enzymes produced are subsequently extracted from the *koji* using brine solutions. *Koji* fermentations are carried out in East Asia at a small home scale, as well as in the large-scale industrial manufacture of soy sauce and rice wine (11). Although mycotoxin-producing molds such as *Aspergillus flavus* and *Asp. parasitious* occur in *koji* as natural contaminations, they have not been observed to produce aflatoxins under the given conditions.

The principle of fungal solid-substrate fermentation may be used to prepare enzyme concentrations for conversion of starch, detoxification of cyanogenic glycosides, and other applications.

DRY MATTER BALANCE

Food fermentation is advantageously used for food preservation and to obtain desirable flavor and digestibility. However, some processes are rather wasteful. For instance, prolonged soaking and microbial respiration of organic matter may lead to considerable losses of valuable raw material dry matter. Examples can be found in the traditional process of *ogi* manufacture (fermented maize cake) and the *tempe* process, during which up to 30 percent of the raw material may be lost by leaching during soaking steps. Encouraging research has been carried out by Banigo et al. (12) in the field of Nigerian *ogi* manufacture, aimed at reducing these raw material losses by omitting soaking stages. It would certainly be worthwhile to investigate dry matter balances of traditional fermentations with a view to reducing losses of raw material by implementing "dry" instead of "wet" processing.

IMPLEMENTATION

No matter how much research is carried out on improved traditional processes or novel products, the ultimate aim is implementation.

Unfortunately, a wide gap exists between research data published in scientific journals and the practice of food processing. Much attention should be given to the extent of usefulness of new products to the end user. To this effect, not only should the sensory, nutritional, and other quality characteristics of newly developed products or processes be taken into account, but they should also be integrated with sound price calculations, market surveys, and extension efforts. Only a competitive process has good chances of being implemented.

In conclusion, the importance of a business-oriented approach and close contact between researchers and food processors, working together toward mutual benefit, must be stressed.

REFERENCES

1. Hesseltine, C. W. 1981. Future of fermented foods. Process Biochemistry 16:2–13.

2. Bruinsma, D. H., and M. J. R. Nout. 1990. Choice of technology in food processing for rural development. Paper presented at the symposium "Technology and Rural Change in Sub-Saharan Africa," Sussex University, Brighton, U.K., Sept. 27–30, 1989. In: Rural Households in Emerging Societies: Technology and Change in Sub-Saharan Africa. M. Haswell, and D. Hunt (Eds.). New York: Berg Publishers.

3. Nout, M. J. R. 1990. Fermentation of infant food. Food Laboratory News 6(2)20:10–12.

4. Spicher, G. 1986. Sour dough fermentation. Chemie Mikrobiologie Technologie der Lebensmittel 10(3/4):65–77.

5. Nout, M. J. R., and F. M. Rombouts. 1990. Recent developments in *tempe* research. Journal of Applied Bacteriology 69(5):609–633.

6. Nout, M. J. R. 1991. Ecology of accelerated natural lactic fermentation of sorghum-based infant food formulas. International Journal of Food Microbiology 12(2/3):217–224.

7. Mensah, P., A. M. Tomkins, B. S. Drasar, and T. J. Harrison. 1991. Antimicrobial effect of fermented Ghanaian maize dough. Journal of Applied Bacteriology 70(3):203–210.

8. Hesseltine, C. W., R. Rogers, and F. G. Winarno. 1988. Microbiological studies on amylolytic Oriental fermentation starters. Mycopathologia 101(3):141–155.

9. Ikediobi, C. O., and E. Onyike. 1982. The use of linamarase in gari production. Process Biochemistry 17:2–5.

10. Nout, M. J. R., and B. J. Davies. 1982. Malting characteristics of finger millet, sorghum and barley. Journal of the Institute of Brewing 88:157–163.

11. Fukushima, D. 1989. Industrialization of fermented soy sauce production centering around Japanese shoyu. Pp. 1–88 in: Industrialization of Indigenous Fermented Foods. K. H. Steinkraus (Ed.). New York: Marcel Dekker, Inc.

12. Banigo, E. O. I., J. M. de Man, and C. L. Duitschaever. 1974. Utilization of high-lysine corn for the manufacture of *ogi* using a new, improved processing system. Cereal Chemistry 51:559–572.

2

Genetic Improvement of Microbial Starter Cultures

Susan K. Harlander

Fermentation has been used for preserving food for hundreds of years and virtually every culture has, as part of its diet, a variety of fermented milk, meat, vegetable, fruit, or cereal products. Microorganisms, including bacteria, yeasts, and mold, produce a wide range of metabolic end products that function as preservatives, texturizers, stabilizers, and flavoring and coloring agents. Several traditional and nontraditional methods have been used to improve metabolic properties of food fermentation microorganisms. These include mutation and selection techniques; the use of natural gene transfer methods such as transduction, conjugation and transformation; and, more recently, genetic engineering. These techniques will be briefly reviewed with emphasis on the advantages and disadvantages of each method for genetic improvement of microorganisms used in food fermentations.

TRADITIONAL GENETIC IMPROVEMENT STRATEGIES

Mutation and Selection

In nature, mutations (changes in the chromosome of an organism) occur spontaneously at very low rates (one mutational event in every 10^6 to 10^7 cells per generation. These mutations occur at random throughout the chromosome, and a spontaneous mutation in a metabolic pathway of interest for food fermentations would be an extremely rare event. The mutation rate can be dramatically increased by exposure of microorganisms to mutagenic agents, such as ultraviolet light or various chemicals, which induce changes in the deoxyribonucleic acid (DNA) of host cells. Mutation rates can be increased to one mutational event in every 10^1 or 10^2 cells per generation for auxotrophic mutants, and one in 10^3 to 10^5 for the isolation of improved secondary metabolite producers. A method of selection is critical for effective screening of mutants as several thousand individual isolates may need to be evaluated to find one strain with improved activity in the property of interest.

Mutation and selection techniques have been used to improve the metabolic properties of microbial starter cultures used for food fermentations; however, there are severe limitations with this method. Mutagenic agents cause random mutations, thus specificity and precision are not possible. Potentially deleterious undetected mutations can occur, since selection systems may be geared for only the mutation of interest. Additionally, traditional mutation procedures are extremely costly and time-consuming and there is no opportunity to expand the gene pool. In spite of these limitations, mutation and selection techniques have been used extensively to improve industrially important microorganisms and, in some cases, yields of greater than 100-times the normal production level of bacterial secondary metabolites have been achieved.

Natural Gene Transfer Methods

The discovery of natural gene transfer systems in bacteria has greatly facilitated the understanding of the genetics of microbial starter cultures and in some cases has been used for strain improvement. Genetic exchange in bacteria can occur naturally by three different mechanisms: transduction, conjugation, and transformation.

Transduction

Transduction involves genetic exchange mediated by a bacterial virus (bacteriophage). The bacteriophage acquires a portion of the chromosome or plasmid from the host strains and transfers it to a recipient during subsequent viral infection. Although transduction has been exploited for the development of a highly efficient gene transfer system in the gram-negative organism *Escherichia coli,* it has not been used extensively for improving microorganisms used in food fermentations. In general, transduction efficiencies are low and gene transfer is not always possible between unrelated strains, limiting the usefulness of the technique for strain improvement. In addition, bacteriophage have not been isolated and are not well characterized for most strains.

Conjugation

Conjugation, or bacterial mating, is a natural gene transfer system that requires close physical contact between donors and recipients and is responsible for the dissemination of plasmids in nature. Numerous genera of bacteria harbor plasmid DNA. In most cases, these plasmids are cryptic (the functions encoded are not known), but in some cases important metabolic traits are encoded by plasmid DNA. If these plasmids are also self-transmissible or mobilizable, they can be transferred to recipient strains. Once introduced into a new strain, the properties encoded by the plasmid can be expressed in the recipient. The lactic acid bacteria naturally contain from one to more than ten distinct plasmids, and metabolically important traits, including lactose-fermenting ability, bacteriophage resistance, and bacteriocin production, have been linked to plasmid DNA. Conjugation has been used to transfer these plasmids into recipient strains for the construction of genetically improved commercial dairy starter cultures.

There are some limitations in the application of conjugation for strain improvement. To exploit the use of conjugative improvement requires an understanding of plasmid biology and, in many cases, few conjugative plasmids encoding genes of interest have been identified or sufficiently characterized. Conjugation efficiencies vary widely and not all strains are able to serve as recipients for conjugation. Moreover, there is no opportunity to expand the gene pool beyond those plasmids already present in the species.

Transformation

Certain microorganisms are able to take up naked DNA present in the surrounding medium. This process is called transformation and this gene transfer process is limited to strains that are naturally competent. Competence-dependent transformation is limited to a few, primarily pathogenic, genera, and has not been used extensively for genetic improvement of microbial starter cultures. For many species of bacteria, the thick peptidoglycan layer present in gram-positive cell walls is considered a potential barrier to DNA uptake. Methods have been developed for enzymatic removal of the cell wall to create protoplasts. In the presence of polyethylene glycol, DNA uptake by protoplasts is facilitated. If maintained under osmotically stabilized conditions, transformed protoplasts regenerate cell walls and express the transformed DNA. Protoplast transformation procedures have been developed for some of the lactic acid bacteria; however, the procedures are tedious and time-consuming, and frequently parameters must be optimized for each strain. Transformation efficiencies are often low and highly variable, limiting the application of the technique for strain improvement.

Electroporation

The above mentioned gene transfer systems have become less popular since the advent of electroporation, a technique involving the application of high-voltage electric pulses of short duration to induce the formation of transient pores in cell walls and membranes. Under appropriate conditions, DNA present in the surrounding medium may enter through the pores. Electroporation is the method of choice for strains that are recalcitrant to other gene transfer techniques; although optimization of several parameters (e.g., cell preparation conditions, voltage and duration of the pulse, regeneration conditions, etc.) is still required.

GENETIC ENGINEERING

Genetic engineering provides an alternative method for improving microbial starter cultures. This rapidly expanding area of technology provides methods for the isolation and transfer of single genes in a precise, controllable, and expedient manner. Genes that code for specific desirable traits can be derived from virtually any living organism (plant, animal, microbe, or virus). Genetic engineering is revolutionizing the science of strain improvement and is destined to have a major impact on the food fermentation industry.

Although much of the microbial genetic engineering research since the advent of recombinant DNA technology in the early 1970s has focused on the gram-negative bacterium *Escherichia coli*, significant progress has been made with the lactic acid bacteria and yeast. Appropriate hosts have been identified, multifunctional cloning vectors have been constructed, and reliable, high-efficiency gene transfer procedures have been developed. Further, the structural and functional properties, as well as the expression in host strains, of several important genes have been reported. Engineered bacteria, yeast, and molds could also be used for the production of other products, including food additives and ingredients, processing aids such as enzymes, and pharmaceuticals.

Prerequisites

Metabolism and Biochemistry of the Host

A necessary prerequisite for the application of genetic engineering to any microorganism is a fundamental understanding of the metabolism and biochemistry of the strain of interest. Although for hundreds of years the metabolic potential of microbial starter cultures has been exploited, in many cases little is known about specific metabolic pathways, the regulation of metabolism, or structural and functional relationships of critical genes involved in metabolism. This information is essential for the design of genetic improvement strategies, as it provides the rationale for selection of desirable gene(s) and assures that once inserted into a new host, the gene(s) will be appropriately expressed and regulated as predicted.

Transformable Hosts

Plasmid-free, genetically characterized and highly transformable hosts, coupled with multifunctional expression vectors, provide the necessary tools for transfer, maintenance, and optimal expression of cloned DNA in microbial starter cultures. Many microbial starter cultures harbor plasmid DNA, and although most plasmids remain cryptic, resident plasmids interfere with identification of plasmid-containing transformants. Use of plasmid-free hosts also eliminates plasmid incompatibility problems and the possibility of cointegrate formation between transforming and endogenous plasmids. It is important to note that plasmid-free strains are used for the development of model systems; however, ultimately it will be necessary to engineer commercial strains.

Vector Systems

A vector can be defined as a vehicle for transferring DNA from one strain to another. Plasmids are frequently used for this purpose because they are small autonomously replicating circular DNA forms that are stable and relatively easy to isolate, characterize, and manipulate in the laboratory. Native plasmids do not naturally possess all of the desirable features of a vector (e.g., multiple cloning sites, selectable marker(s), ability to replicate in several hosts, and so forth). Therefore, genetic engineering is frequently used to construct multifunctional cloning vectors. Although antibiotic resistance markers greatly facilitate genetic engineering in microbial systems, vectors derived solely from food-grade organisms may be critical in obtaining regulatory approval for use of the organisms, as antibiotic resistance determinants may not be acceptable in food systems.

An alternative vector strategy involves the development of linear fragments of DNA that are capable of integrating into the host chromosome via homologous recombination. Although transformation frequencies are very low, the advantage of the integrative vector is that transformed genetic information is targeted to the chromosome where it will be more stably maintained. Insertion sequences (IS elements) naturally present in the chromosome that can transpose chromosomal DNA to plasmids could be used as an alternative strategy for developing integrative vectors for some strains of lactic acid bacteria.

Efficient Gene Transfer Systems

Once gene(s) have been identified and cloned into the appropriate vector in the test tube, they must be introduced into a viable host. Since the recombinant DNA is a naked DNA molecule, gene transfer systems based on protoplast transformation and electroporation are most applicable in genetic engineering experiments. High transformation efficiencies (greater than 10^4 to 10^5 transformants per kilogram of DNA) greatly facilitate screening and identification of appropriate transformants. Electroporation is the transformation procedure of choice for most microbial strains.

Expression Systems

Transfer of structural genes to a new host using genetic engineering does not guarantee that the genes will be expressed. To optimize expression of cloned genes, efficient promoters, ribosome-binding sites, and terminators must be isolated, characterized, and cloned along with the gene(s) of interest. Identification of signal sequences essential for secretion of proteins outside the cell may be useful for situations where microbial starter cultures are used to produce high-value food ingredients and processing aids. Secretion into the medium greatly facilitates purification of such substances.

Properties of Interest

Several properties could be enhanced using genetic engineering. For example, bacteriocins are natural proteins produced by certain bacteria that inhibit the growth of other often closely related bacteria. In some cases, these antimicrobial agents are antagonistic to pathogens and spoilage organisms commonly found as contaminants in fermented foods. Transfer of bacteriocin production to microbial starter cultures could improve the safety of fermented products.

Acid production is one of the primary functions of lactobacilli during fermentation. Increasing the number of copies of the genes that code for the enzymes involved in acid production might increase the rate of acid production, ensuring that the starter will dominate the fermentation and rapidly destroy less-aciduric competitors.

Certain enzymes are critical for proper development of flavor and texture of fermented foods. For example, lactococcal proteases slowly released within the curd are responsible for the tart flavor and crumbly texture of aged cheddar cheese. Cloning of additional copies of specific proteases involved in ripening could greatly accelerate the process.

An engineered *Saccharomyces cerevisiae* (baker's yeast), which is more efficient in leavening of bread, has been approved for use in the

United Kingdom and is the first strain to attain regulatory approval. This strain produces elevated levels of two enzymes, maltose permease and maltase, involved in starch degradation.

Limitations

There are a number of issues that must be resolved before genetically engineered starter cultures could be used in food. Engineered strains will need to be approved for use by appropriate regulatory agencies. To date, no engineered organisms have been approved in the United States, and specific criteria for approval have not been established by the Food and Drug Administration.

The public must be assured that the products of biotechnology are safe for consumption. If consumers have the perception that the products are not safe, the technology will not be utilized. Although genetic engineering is probably safer and more precise than strain-improvement methods used in the past, most U.S. consumers are not aware of the role of bacteria in fermented foods and do not have a fundamental understanding of recombinant DNA technology, and they may be unwilling to accept the technology. This may be less of a problem in developing countries where improved microbial starter cultures could provide significantly safer and more nutritious foods with longer shelf life and higher quality.

Another limitation is that genetic improvement of microbial starter cultures requires sophisticated equipment and expensive biological materials that may not be available in developing countries. Where equipment and materials are available in industrialized countries, there may be little incentive for researchers to improve strains that would probably not be used in their own countries.

Genetic improvement of microbial starter cultures is most appropriate for those fermentations that rely solely or primarily on one microorganism. In many cases, our knowledge about the fermentation is limited, making selection of the target strain very difficult. Since many food fermentation processes are complex and involve several microorganisms, genetic improvement of just one of the organisms may not improve the overall product.

3

Sudan's Fermented Food Heritage

Hamid A. Dirar

If we accept the idea that Africa is the birthplace of Man, it would seem logical that the first human or humanoid to consume a fermented food would have lived there. That fermented product could have been a piece of meat or some kind of berry picked up or stored by a hunter-gatherer. Later, and after those early men, or rather women, developed the taste for such goods they began to intentionally store fresh food items to undergo spontaneous fermentation.

Should this be the case, one would expect to find in Africa today a diverse array of fermented food products. Unfortunately, we know very little about African fermented foods because no genuine attempt has been made by any African scientist to document all the fermented foods of his or her country.

For at least one African country, the Sudan, I set out 6 years ago to collect, confirm, reconfirm, sift, and classify information on all fermented foods in the country. The major source of information was the elderly rural women of Sudan. The list of fermented foods and beverages, which now includes 60 different items, will make the basis for a book that should be ready for publication within a year. In the following sections I discuss some of the important aspects that came out of this personal initiative, which was not in any way sponsored by any agency, except perhaps some help from Band Aid of Britain.

FERMENTED FOODS

The Sudanese seem to bring just about anything edible or barely edible to the forge of the microbe, to the extent that one could confidently say: food in Sudan is fermented. The raw materials to be fermented include the better-known products such as sorghum, millet, milk, fish, and meat. Also, a number of unorthodox raw materials are fermented: bones, hides, skins, hooves, gall bladder, fat, intestines, caterpillars, locusts, frogs, and cow urine.

The bulk of these foods is poured into the bowl of sorghum porridge, being either a sorghum (or millet) staple or its sauce and relish. The few remaining ones are alcoholic or nonalcoholic beverages, the most important of which are prepared from sorghum. In other words, every fermented food item orbits around the sorghum grain.

Sorghum-Based Foods

Sorghum fermented foods and drinks are the most sophisticated and are prepared by the most complicated procedures. Compared with similar sorghum products of Africa and indeed of the whole world, the Sudan's sorghum products stand out as unique in many respects:

- The Sudan seems to have the greatest number of fermented sorghum products. There are about 30 such products that are basically different from one another.
- There is a wide use of sorghum malt in the preparation of food and drink. Throughout Africa sorghum malt is more commonly used in the preparation of beers. In Sudan, however, while malt is used in three major beer types, it is also used to make some seven solid food products. This situation does not seem to hold true for other African countries, judging by the literature.
- The making of bread-type foods from sorghum is not common in Africa. The Sudan, however, has about 12 sorghum breads (discs, sheets, flakes). Close scrutiny of these breads reveals an influence from the Middle East; some of these breads carry names and are prepared by methods used for similar products in the Arab World.
- A comparison of the procedures followed in the preparation of some sorghum food products in Sudan with procedures for making similar products in other African countries suggests that the art of making these products traveled from Sudan to West Africa and perhaps to East Africa, too. In some cases the product travelled carrying the same Arabic-Sudanese name.

This suggests that sorghum food culture is more ancient than in other areas of Africa, and this food evidence may be taken to strengthen previous hypotheses that the origin of sorghum domestication is somewhere in northeast Africa.

Dairy Products

The most common fermented milk product of Sudan is *rob*. Milk is fermented overnight, and the resulting sour milk is churned to give butter; the remaining buttermilk is *rob*. The principal aim behind *rob* production is the need to facilitate the extraction of butter from the milk. The butter (*furssah*) is later boiled to give butter oil or ghee, which can be stored for use in the lean season. *Rob* production is in the hands of animal-owning nomadic tribes, and the bulk of it is produced during the rainy season (July-October). Huge amounts of *rob* are thrown away during this season as useless after the butter has been removed. Some women, however, allow the souring process to proceed further after butter extraction until the curd is separated from the whey. They then collect the curd and sun dry it to give a kind of granular cheese called *kush-kush* that is turned into sauce for sorghum porridge in later months.

Another kind of sour milk is fermented camel milk, called *gariss*. This is probably the only fermented food product invented by men. *Gariss* is prepared by camel boys who depend on it as their major nourishment when they roam with their herds into remote areas. The milk is fermented in a skin bag hitched to the saddle of a camel that is allowed to go about its business as usual—grazing, sleeping, walking, trotting, etc. This product, unlike *rob*, is fermented for consumption and no butter is removed from it.

A third indigenous dairy product is *biruni,* also called *leben-gedim,* which is a fermented unchurned milk ripened for up to 10 years! A related product, but not ripened, is *mish*, which is made by prolonged fermentation to the extent that maggots thrive in it. The product is consumed whole, with the maggots included. These two products are closely related to Egyptian *mish* (1).

Dairy products that have entered the Sudan from Egypt within the last century are *jibnabeida* (white cheese), *zabadi* (yogurt), and black cumin-flavored *mish*. These products are strictly confined to urban communities, where the Egyptian influence is more strongly felt.

Fish Products

Southeast Asia takes all the fame in the literature concerning the production of fermented fish products. But if one sorts out all the various products of that corner of the world carrying a confusing array of names, one finds that the products boil down to four major categories: sauces, pastes, dried fish, and whole salted fish. These four types of fermented fish products are also found in the Sudan, only they are all prepared from freshwater Nile fish. This situation has not been reported for other African or Arab countries. The Sudanese fish products include *kejeik* (large sun-dried split fish); *fessiekh* (salted fermented whole tiger fish); *mindeshi* (pounded small fish paste, fermented, and may be dried later); and *terkin* or *meluha* (fermented fish sauce or paste—not dried).

Meat Products

While some urban people in Sudan make very thin strips of red beef and dry them in the sun to give *shermout,* the traditional rural product is a truly fermented one. Thick strips of fat-bearing meat are hung on a rope indoors and left to undergo fermentation and slow drying to give a proteolytic product, *shermout.*

The Sudanese also ferment the sheath of fat surrounding the stomach to give the strongest-smelling product of all, *miriss.* Others ferment the small intestines to give *musran.* The clean small intestines may also first be sun dried together with strips of the lungs, heart, kidneys, liver, etc., and then all pounded together and mixed with some potash and molded into a fist-sized ball and allowed to slowly ferment and dry, to give *twini-digla.* The large intestine is cleaned and stuffed with fat and hung to ferment and dry for a month, to give the sausage called *skin.*

Beirta is prepared from he-goat meat. Small pieces of muscle meat, lungs, kidneys, liver, heart, etc., are mixed with milk and salt, packed into a clay pot, and allowed to undergo some sort of pickling, presumably.

Um-tibey is best prepared from gazelle's meat. The rumen is carefully emptied and then stuffed with the vertebrae of the neck, cut-up heart, kidneys, liver, etc. The rumen is next tied and hung high to undergo fermentation. The whole thing may then be cooked by burying it in hot ashes and embers.

Fresh bones may be fermented in a number of ways. The large bones, with pieces of attached meat and tendons, may simply be thrown on a thatched roof to ferment slowly for weeks or even months to give the product called *adum* (bone). The meshy ball bone endings of the ball and socket joints may be pounded fresh and fermented into a paste called *dodery.* The vertebrae of the backbone may be chopped into smaller pieces that are sun dried, pounded with stones, mixed with a little water and salt, molded into a ball, and allowed to ferment and dry to give *kaidu-digla* (bone ball).

The fresh hide, skin, or hoof may be buried in mud or moist ash to undergo fermentation. The fermented product can then be cut into strips or pieces and sun dried and stored. The gall bladder is removed full with its gall juice. Some sorghum flour or grains are added to the juice to absorb it and then hung to undergo slow drying. The product, *itaga,* is later pounded into a sort of spice usually consumed with fatty meat dishes.

Vegetable Products

A number of fermented vegetable products are produced in rural Sudan. Interestingly, these products can be grouped into either meat substitutes or sour milk (*rob*) substitutes, the two major flavors of sauces in the country. *Kawal* (2,3) is the major meat substitute. It is a strong-smelling product derived by fermentation of the pounded green leaves of the wild legume *Cassia obtusifolia,* which grows during the rainy season. The product is used in the preparation of sauces to completely replace meat or for use as a meat extender. Its protein is of high quality, rich in the sulfur amino acids. *Furundu,* a similar meat substitute, is prepared from the seeds of red sorrel *Hibiscus sabdariffa*. *Sigda* is another meat substitute and is prepared by fermentation of sesame oilseed presscake. All these products are dried after fermentation in the form of hard, irregular, small balls and may keep for a year or so. Other ill-defined but related products are *kerjigil* (from a mixture of pumpkins, sesame, and cowpea) and *teshnuti* (from okra seeds).

Sour milk (*rob*) substitutes are made from oil-bearing seeds in a manner analogous to the use of soybeans to give dairy product analogs. *Rob-heb* is made from the seeds of the watermelon. *Rob-ful* is made from peanuts. In either case the seeds are pounded into a paste that is allowed to undergo a souring fermentation. When mixed with water and turned into sauce the product has the color (off white) and taste (sour) of the sour milk sauce called *mulah-rob*. A related product is *um-zummatah*, obtained by the souring fermentation of watermelon juice. The same name is sometimes given to the sour steep water, also called *mayat-aish,* of fermented whole sorghum or millet grain.

Alcoholic Products

Opaque beers are commonly brewed in Africa but procedures vary. The brewing of *merissa* in Sudan is probably the most complicated and advanced of all (4,5). The unique features of this brewing method include the use of only a small amount (5 percent) of sorghum malt as an enzyme preparation, rather than a substrate. Malt constitutes 25 to 100 percent of the substrate in the brewing of most African and European beers. Another unique feature is the use of a caramelized sorghum product, called *surij*, in the process. Third, there is a special starter activation step during the process that is lacking from other African brewing procedures. Also, the brewer women seem to be aware of the properties of enzymes and microbes as well as those of the acids produced during fermentation. This explains the unique treatment of the substrate, where parts of it are half cooked, others fully cooked, and yet others overcooked to meet enzyme requirements for a mixture of raw and gelatinized starch and to effect sterilization of products when needed. The *merissa* process has been well recognized as a complex process that deserves further investigation.

Clear beers are not common in Africa, and the literature gives reports only on *otika* of Nigeria and *amgba* of Cameroon (6,7). The Sudan has a clear sorghum (or millet) beer called *assaliya* (or *um-bilbil*). A look at the production of these three beers reveals that the *assaliya* process, involving some 40 steps, is far more complicated than the *otika* or *amgba* procedures, which involve fewer than 20 steps. It is suggested that the art of brewing clear beers traveled to West Africa from Sudan. *Amgba* of Cameroon is even called *bilbil.*

In Sudan there are perhaps 30 to 50 opaque beer types with different but related brewing methods. The area seems to be a center of diversity of sorghum beers, and perhaps the art of brewing of opaque beers traveled to East Africa from this region.

The traditional wines of Sudan are the date wines. The palm wine of West Africa is not known in Sudan—nor is *lagmi,* the wine obtained by fermentation of the sap of the date palm as practiced in northwest Africa. Only the fruit of the date palm is fermented in the Sudan, and the bulk of wines thus made are produced and consumed in the Northern Province where most of the date palms exist. At least 10 different date wines are produced, the most important of which are *sherbot*, *nebit*, and *dakkai* (8).

In the Southern Sudan a kind of mead is produced by fermentation of diluted wild bee's honey. The product, called *duma*, is primed by a specially prepared starter culture called *duma*-grains (*iyal-duma*).

FERMENTED FOODS AND SURVIVAL STRATEGIES

A careful examination of fermented food products of Sudan would immediately suggest a close link between food fermentation and food shortage in this part of the world. First, about 80 percent of these foods, particularly the marginal ones using bones, intestines, fat, etc., are found in western Sudan in the Kordofan and Darfur regions, the traditional famine areas. Second, most of the foods are preserved by both fermentation and drying, which means that they are intended for long storage and that food shortages or even famine are anticipated. In other words, the inventors of such foods have the experience of repeated famines.

Further, practically all fermented sauce ingredients are produced during the late months of the rainy season, which shows that, unless a person secures all of his or her food requirements from this short season, he or she will probably suffer greatly in the remaining 9 months of the year. The harsh environment has actually dictated the need to ferment and dry anything that might prevent starvation. To live on the edge of the desert must have been a great force in sharpening the sense for survival and creativity.

The strong link between many fermented foods and food shortages is also revealed by the fact that if a family became rich it would drop a number of fermented foods from its menu, not because of social pressure but because there was no longer any need for them now that ample supplies of meat, milk, poultry, etc., were available. Poor people who ferment bones, hides, locusts, etc., do so not because they relish these foods but because it is part of the coping strategy they follow to deal with the vagaries of a capricious environment.

The first victims of any famine are the children, among whom death exacts a great toll. Babies and children die in the laps of women more than they do in the laps of men. Maternal compassion must be the greatest impetus behind the rural woman's desperate attempts to save her child that propel her to look for an insect, a piece of hide, a frog, or a bone as savior. Many fermented foods are thus famine foods, and rural women must be credited with their invention. These women must have saved thousands of children from certain death during famines. Their vital role must be recognized and hailed.

BIOTECHNOLOGY AND FERMENTED FOODS

This relationship has not been discussed widely in the literature. One can imagine, however, that biotechnology can be of help in the improvement of fermented foods at three levels:

- Raw materials. Fermented foods are produced from either animal or plant starting materials, and the availability of these substrates will of course aid in the production of fermented foods. Biotechnological methods to improve animal and plant production have been dealt with by experts in those fields on many occasions.

  Only a special reminder should be made not to neglect certain wild plants and marginalized crops—the so-called lost crops of Africa (e.g., sorrel and okra). Attempts to restore the forest cover should give some attention to trees that bear fruits used during famines or even trees that host caterpillars.
- Fermentation engineering. Recent developments in biotechnology have given rise to great innovations in bioreactor designs. Most of these designs deal with liquid reaction media, but it should not be forgotten that a great number of fermented foods are produced through a solid-substrate fermentation in which the fermenting paste is frequently hand mixed. Bioreactors to simulate such a process are needed for the modernization of such traditional fermented foods.
- Microbiology and enzymology. There are many opportunities for biotechnological innovations in the microbiology of fermented foods.

First, all the microorganisms involved should be isolated, characterized, and preserved as a germplasm collection. Second, the metabolic role of each of the strains involved should be clearly identified, and their full potential, even in other fields of biotechnology, should be studied. The powerful technique of monoclonal antibodies for the characterization of different strains of the same species can be of great help in this area.

Many of these organisms have the enzyme complement to produce vitamins and amino acids in fermented foods. This potential can be improved through the technique of recombinant DNA technology to produce strains that are capable of producing and releasing the required amino acid or vitamin into the food.

To avoid food losses due to spoilage-causing organisms and to avoid possible development of food-poisoning microbes, it is possible to genetically engineer a strain required for a process as a pure culture. Such a strain may bring about all the changes required in the food and grow at a convenient temperature.

REFERENCES

1. Abdel-Malek, Y. 1978. Traditional Egyptian dairy fermentations. Global Impacts of Applied Microbiology 5:198-208.

2. Dirar, H. A. 1984. *Kawal*, a meat substitute from fermented *Cassia obtusifolia* leaves. Economic Botany 38:342-349.

3. Dirar, H. A., D. B. Harper, and M. A. Collins. 1985. Biochemical and microbiological studies on *kawal*, a meat substitute derived by fermentation of *Cassia obtusifolia* leaves. Journal of the Science of Food and Agriculture 36:881-892.

4. Dirar, H. A. 1976. The art and science of *merissa* fermentation. Sudan Notes and Records 57:115-129.

5. Dirar, H. A. 1978. A microbiological study of Sudanese *merissa* brewing. Journal of Food Science 43:1683-1686.

6. Ogundiwin, J. O. 1977. Brewing *otika* ale from guinea corn in Nigeria. Brewing and Distilling International 7(6):40-41.

7. Chevassus-Agnes, S., J. C. Favier, and A. Joseph. 1976. Technologie traditionelle et valeur nutritive des "bieres" do sorgho du Cameroon. Cahier de Nutrition et de Dietetique 11(2):89-104.

8. Ali, M. Z., and H. A. Dirar. 1984. A microbiological study of Sudanese date wines. Journal of Food Science 49:459-460, 467.

4

Lesser-Known Fermented Plant Foods

Kofi E. Aidoo

In many parts of the world, fermented foods form an important part of the diet. These foods are made from plant and animal materials in which bacteria, yeasts, and molds play an important role by modifying the material physically, nutritionally, and organoleptically.

Fermented plant foods may be classified into groups as (a) those made from cereal grains (maize, sorghum, millet, rice, wheat), such as *pozol* (Mexico), *kenkey, ogi,* and *injera* (Africa); (b) those made from pulses, nuts, and other seeds, such as *ontjom* (Indonesia) and *dawadawa* (Savannah Africa); (c) those from tubers (cassava, aroids, potatoes), such as *gari* (Africa) and *farinha puba* (Brazil, Peru, and Ecuador); (d) those from fruits and vegetables, such as *gundruk* (Nepal) and *kimchi* (Korea, East Asia); and (e) beverages derived from tree saps, such as *nipa* wine (Far East) and *pulque* (Mexico).

Most traditional fermented plant foods are prepared by processes of solid-substrate fermentation in which the substrate is allowed to ferment either spontaneously (usually African or Latin American processes) or by adding a microbial inoculum (Far East, South Asia, and Southeast Asia).

Cereal grains account for more than 60 percent of food materials used in the preparation of indigenous fermented foods in Africa. Although maize is a comparatively well-researched crop, no significant research has been done on some of the important traditional crops, such as sorghum and millet (1). *Tef* (*Eragrostis tef*), a staple food grain of Ethiopian subsistence farmers, is still relatively less well known.

Many indigenous fermented foods, some of which long predate recognition of the existence of microorganisms, are eaten in various parts of the world. Increasing interest in this field is reflected in the range of publications (2-10). This paper presents information on some of the lesser-known fermented plant foods that are still produced and marketed on a small scale and that serve as a staple diet for millions of people in developing countries.

REGIONAL PERSPECTIVES

Cereals are major staples in many developing countries, and the fermentation of cereal grains to prepare a variety of foods has a long history. Fermented products from maize are usually found in Africa and Central and South America and those from sorghum (guinea corn) and millet in Africa and South Asia. Food fermentations based on rice are practiced in India, China, Southeast Asia, and the Far East, while those from wheat are particularly important in the Middle East, Turkey, and the Far East (11).

Fermented foods from tubers are usually found in Africa, among the Andean Indians and in the South Pacific, and the process of detoxification of the tuber before fermentation is still carried out by soaking in water.

Chica, an alcoholic beverage made from maize in Peru since pre-Hispanic times, also is produced from potato, *oca* (*Oxalis tuberosa*), *arracacha* (*Arracacia xanthorrhiza*), *maca* (*Lepidium evenii*), and other Incan crops that science has almost totally neglected. Although cassava and sweet potatoes provide nourishment for more than 500 million people, only a small proportion of this highly perishable staple crop is used in food fermentations in Africa and Latin America.

Legumes account for a substantial amount of food protein intake in developing countries. Of the total world production of over 58 million metric tons in 1990, developing countries produced 62 percent, together with 54 percent of world nut production (12). Fermented products from legumes are not as popular in Africa or Latin America as in the Far East and South and Southeast Asia, where soybean, for instance, is used extensively in the production of fermented products such as soy sauce, *miso*, and *tempe*, and black gram dhal for the production of *idli* and *dosa*. Fermented seed products, however, are often used as condiments in Savannah Africa.

In the tropics, highly perishable foods such as fruits and vegetables may be preserved as fermented products. Some fermented vegetables provide vitamins, particularly during long cold months in the northern parts of East Asia, and others are consumed as part of traditional family life in Southeast Asia. In Mexico refreshing beverages are prepared from a variety of fruits, including pineapples, apples, and oranges.

PRODUCTS FROM CEREAL GRAINS

Ahai

Ahai is a sweet, malty-tasting beverage brewed from maize in Southern Ghana and is usually served as a welcome drink and at outdoor ceremonies, wakes, and funerals. Whitby (13) has reported that the traditional method of preparing *ahai* is much the same as for *pito*, an acid-alcohol beer brewed from sorghum or millet in West Africa, except that *ahai* is not boiled again after fermentation. So far, no studies have been made on the microbiological, biochemical, and nutritional changes that take place during *ahai* production.

Ting

Ting is a staple food for a large proportion of the population of Botswana. It is prepared from maize by natural fermentation. In other regions it is prepared from sorghum or millet. Moss et al. (14) made an extensive study of *ting* fermentation and noted that the success of the fermentation depends on a number of factors, among which temperature is very important.

The microbiology of *ting* fermentation is well documented, but further studies need to be carried out, particularly on the nutritional value. *Ting* may be similar, nutritionally, to other acid-fermented cereal gruels like *kenkey* (West Africa), *kisra* (Sudan), and *pozol* (Mexico).

Maasa

Maasa is a snack food made from millet or sorghum and is very popular in Savannah Africa, particularly during Ramadan. The method of preparation of *maasa* has been reported (9), but there is no information on the microbiology and biochemistry of this fermented product.

There are hundreds of fermented products from cereal grains in the tropical regions of the world that require extensive studies on methods of preparation and biochemical, microbial, and nutritional changes. These include the West African *fura* or *fula, jamin-bang* of the Kaingang Indians of Brazil, and the Maori's *kaanja-kopuwai*, a process of fermenting maize in water prior to eating. The Maoris claim *kaanja-kopuwai* is health giving, and many of the older people attribute their age to this part of their diet.

PRODUCTS FROM ROOT TUBERS

Farinha puba

Farinha puba is a coarse flour made from cassava and is found in the Amazonian regions of Brazil, Peru, and Ecuador. Woolfe and Woolfe (15) presented an outline on the preparation of *Farinha puba*, which is also known as *farinha de mandioca* in Brazil. They noted that the technology was exported to West Africa in the nineteenth century and presumably adapted locally to give the *gari* process. *Gari*, a popular West African staple food that is also eaten in other tropical African countries, is prepared by fermenting cassava; details of improved methods of production are given by Steinkraus et al. (6).

The processes involved in the production of *farinha puba* and *gari* are similar, but unlike *gari* very little information has been published on the methods of production and on the microbiology, nutritional values, and toxicological problems of *farinha puba*. It has been reported that cassava fermentation as practiced in Africa, Asia, and Latin America (16) is an unreliable detoxification method, and the process further reduces the already low protein content. Other studies have shown that cassava fermentation for *gari* production does not totally eliminate the cyanide content but reduces it by at least 65 percent (17,18).

Fatalities from cassava poisoning appear to be rare, but long-term toxic effects, (e.g., goiter and cretinism) in cassava-consuming populations may be more serious, especially in the Amazon, where the pressed-out juices are used for making soups and stews (15). In Africa the pressed-out juice is often used for the production of cassava starch for laundry purposes. The use of pure microbial cultures under controlled fermentation conditions might bring about not only complete hydrolysis of the poisonous glycoside but also an enhanced fermentation process.

Kokonte

Kokonte, another important cassava-based staple, is eaten by millions of people in Savannah Africa. Like many other fermented foods, *kokonte* (Ghana) is known by various names such as *ilafun* (Nigeria) and *icingwadal* (East Africa). The method of preparation of *kokonte* has been reported, but further studies need to be done, particularly on microflora and production of mycotoxins during fermentation (19,20).

Masato (masata)

Masato, or cassava beer, is an alcoholic beverage produced from cassava in the Amazon. It has an alcohol content of 6 to 12 percent by volume and is offered to guests as a sign of hospitality. It is considered an offense to refuse a drink (15). In Brazil it is called *kaschiri* and in Mozambique *masata*. Preparation of *masato* is similar to that of *chica* by the Andean Indians. As a first step of fermentation, cassava is chewed and spat out by women. In Mozambique women chew the yucca plant to produce a similar product.

So far, no scientific account of the *masato* fermentation process has been published. Studies on improving the traditional methods of production are necessary to save this ancient art of the Andean Indians from extinction.

Chuno

Chuno is a food product from potato prepared by the inhabitants of the high Andes of Peru, Chile, Ecuador, Colombia, and Bolivia. An outline of the method of production has been reported, but the microorganisms involved in the fermentation are still not known (9).

The Incan anu (*Tropaecolum tuberssum*) is a tuber that must be fermented before being eaten baked, fried, or added to stew (21). The crop is cultivated in Colombia, Peru, and Bolivia and is also grown as a flowering ornament in Britain and the United States. The fermentation involved during "curing" has not been reported.

PRODUCTS FROM LEGUMES, PULSES, AND OTHER SEEDS

In Savannah Africa, fermented products from legumes and other seeds are important food condiments and are generally strong smelling. Quite often seeds that are used for fermentation are inedible in their raw unfermented state. Fermentation of the West and Central African *iru* or *dawadawa* is similar to the Japanese *natto*, and there is adequate literature on the preparation, biochemistry, microbiology, and industrialization of *iru*. Other indigenous products that are receiving some attention include *ugba* (African oil bean seed), *ogiri* (seeds of watermelon), *ogiri-igbo* (castor oil seed), and *ogiri-nwan* (fluted pumpkin beans).

Lupins (*Lupinus mutabilis*), which are native to the Andes, contain bitter alkaloids and can cause toxicity problems. Lupin seeds are debittered by soaking them in running water, a process similar to the Maoris' process for corn fermentation and the Ichunol methods of Peru and Bolivia. So far, no report has been published on the debittering of lupins by fermentation, but the soaking may involve some fermentation.

Kenima is a Nepalese fermented product from legumes. There is no published information on the method of preparation, microbiology, and nutritional value.

PRODUCTS FROM FRUITS AND VEGETABLES

Colonche is a sweet fizzy beverage produced in Mexico by fermenting the juice of tunas (fruits of the prickly pear cacti, mainly *Opuntia* species). *Tepache* is also a refreshing beverage prepared originally from maize but from various fruits and is consumed throughout Mexico.

Although some studies have been made on these products (22), it appears that more work is needed, particularly on the biochemical and nutritional changes that take place during the preparations.

The Nepalese pickle or *gundruk* is a fermented dried vegetable served as a side dish with the main meal and is also used as an appetizer in the bland starchy diet. Several hundred tons of *gundruk* is produced annually, and production is still at the household level. Dietz (23) reported on the method of preparation and the role of *gundruk* in the diet of Nepalese people. It has been found that a disadvantage of the traditional process is loss of 90 percent of the carotenoids. Improved methods and further studies might help reduce vitamin loss.

COMMERCIALIZATION

To industrialize some of these fermented plant foods from traditional processes, extensive studies must be made to determine the essential microorganisms, optimum fermentation conditions, biochemical changes, nutritional profile, and possible toxicological problems associated with certain plant materials or the fermented product itself.

Commercial or large-scale processes for indigenous fermented foods need to be adapted to specific local circumstances. Advantages of industrialization include a product with an extended shelf life, maximum utilization of raw materials, production of important by-products, and bioenrichment or fortification of a product for specific consumers such as special diets, weaning foods and exclusion of or reduction in the levels of mycotoxins. Mycotoxins appear to be a major problem in some fermented products, particularly those of cereal and root tuber origin.

Studies in Japan on *okara*, a by-product of the tofu industry, have shown that fermenting it with *tempe* fungus could result in a product that is useful as a high-fiber, low-energy food material (24).

REFERENCES

1. Cross, M. 1985. Waiting for a green revolution. New Scientist 1486:30.

2. Hesseltine, C. W. 1965. A millennium of fungi, food and fermentation. Mycologia 57:149–197.

3. Hesseltine, C. W. 1983. The future of fermented foods. Nutrition Review 41:293–301.

4. Rose, A. H. 1982. Economic Microbiology. Fermented Foods, Vol. 7, London: Academic Press.

5. Steinkraus, K. H. 1983. Fermented foods, feeds and beverages. Biotechnology Advances 1:31–46.

6. Steinkraus, K. H. 1983. Handbook of Indigenous Fermented Foods. New York: Marcel Dekker.

7. Beuchat, L .R. 1983. Indigenous fermented foods. Pp. 477–528 in: Biotechnology, Vol. 5. H. J. Rehm and G. Reed (Eds.) Weinheim: Verlag Chemie.

8. Wood, B. J. B. 1985. Microbiology of fermented foods. London: Elsevier Applied Science Publishers. Vols. 1 and 2.

9. Campbell-Platt, G. 1987. Fermented Foods of the World: A Dictionary and Guide. London: Butterworths.

10. Berghofer, E. 1987. Use of non-European fermented foods in Austrian market. Ernahrung 11(1):14–22.

11. Hesseltine, C. W. 1979. Some important fermented foods of mid-Asia, the Middle East and Africa. Journal of the American Oil Chemists Society 56:367–374.

12. FAO. 1991. Food and Agriculture Organization Quarterly Bulletin of Statistics 4(1).

13. Whitby, P. 1968. Foods of Ghana. Food Research Institute Report 1:1–31.

14. Moss, M. O., S. F. Mpuchane, and O. M. Murphy, 1984. Ting—a fermented maize meal product of southern Africa. Proceedings of the Institute of Food Science and Technology 17:139–148.

15. Woolfe, M., and J. Woolfe. 1984. Some traditional processed

foods of South America. Proceedings of the Institute of Food Science and Technology (U.K.) 17:131–138.

16. Seneviratne, G. 1985. Making cassava a safer food. Development Forum, UNDESI/DPI and UNU, 13:13.

17. el Tinay, A. H., P. L. Bureng, and E. A. E. Yas. 1984. Hydrocyanic acid levels in fermented cassava. Journal of Food Technology 19:197–202.

18. Ayernor, G. S. 1985. Effect of the retting of cassava on product yield and cyanide detoxification. Journal of Food Technology 20:89–96.

19. Aidoo, K. E. 1986. Lesser-known fermented plant foods. Tropical Science 26:249–258.

20. Aidoo, K. E. 1991. Postharvest storage and preservation of tropical crops. Pp. 747–764 in: Mycotoxin and Animal Foods. J. E Smith and R.S. Henderson, Eds. Boca Raton, Fla.: CRC Press.

21. Vietmeyer, N. 1984. Lost crops of the Incas. Ceres 17(3):37–40.

22. Ulloa, M. 1980. Indigenous fermented beverages of Mexico. Pp. 45–49 in: Global Impact of Applied Microbiology. S. O. Emejuaiwe, O. Ogunbi, and S. O. Sanni Eds., London: Academic Press.

23. Dietz, H. M. 1984. Fermented dried vegetables and their role in nutrition in Nepal. Proceedings of the Institute of Food Service and Technology (U.K.) 17:208–213.

24. Matsuo, M. 1989. Morphological and physicochemical properties and composition of Okara fermented with *Rhizopus oligosporus*. Journal of the Japanese Society of Nutrition and Food Science 42(2):173-178.

25. Uchimura, T., V. V. Garcia, and D. M. Flores. Microbiological studies on fermented rice cake, 'puto' and the application of puto making using cassava flour. Tropical Root Crops: Postharvest Physiology and Processing. I. Uritani and E. D. Reyes (Eds.). Tokyo: Japanese Science Society Press.

26. Mabbett, T. 1991. Local strains make good. African Farming and Food Processing Jan/Feb: 25–26.

5

Lactic Acid Fermentations

Keith H. Steinkraus

Lactic acid bacteria perform an essential role in the preservation and production of wholesome foods. The lactic acid fermentations are generally inexpensive, and often little or no heat is required in their preparation, making them fuel efficient as well. Foods fermented with lactic acid play an important role in feeding the world's population on every continent.

Lactic acid bacteria perform this essential function in preserving and producing a wide range of foods: fermented fresh vegetables such as cabbage (sauerkraut, Korean *kimchi*); cucumbers (pickles); fermented cereal yogurt (Nigerian *ogi*, Kenyan *uji*); sourdough bread and bread-like products made without wheat or rye flours (Indian *idli*, Philippine *puto*); fermented milks (yogurts and cheeses); fermented milk-wheat mixtures (Egyptian *kishk*, Greek *trahanas*); protein-rich vegetable protein meat substitutes (Indonesian *tempe*); amino acid/peptide meat-flavored sauces and pastes produced by fermentation of cereals and legumes (Japanese *miso*, Chinese soy sauce); fermented cereal-fish-shrimp mixtures (Philippine *balao balao* and *burong dalag*); and fermented meats (e.g., salami).

Lactic acid bacteria are generally fastidious on artificial media, but they grow readily in most food substrates and lower the pH rapidly to a point where competing organisms are no longer able to grow. *Leuconostocs* and lactic streptococci generally lower the pH to about 4.0 to 4.5, and some of the lactobacilli and pedicocci to about pH 3.5, before inhibiting their own growth.

In addition to producing lactic acid, lactobacilli also have the ability to produce hydrogen peroxide through oxidation of reduced nicotinamide adenine dinucleotide (NADH) by flavin nucleotide, which reacts rapidly with gaseous oxygen. Flavoproteins, such as glucose oxidase, also generate hydrogen peroxide and produce an antibiotic effect on other organisms that might cause food spoilage; the lactobacilli themselves are relatively resistant to hydrogen peroxide.

Streptococcus lactis produces the polypeptide antibiotic nisin, active against gram-positive organisms, including *S. cremoris*, which in turn produces the antibiotic diplococcin, active against gram-positive organisms such as *S. lactis*. Thus, these two organisms compete in the fermentation of milk products while inhibiting growth of other gram-positive bacteria.

Carbon dioxide produced by heterofermentative lactobacilli also has a preservative effect in foods, resulting, among others, from its flushing action and leading to anaerobiosis if the substrate is properly protected.

Brining and lactic acid fermentation continue to be highly desirable methods of processing and preserving vegetables because they are of low cost, have low energy requirements for both processing and preparing foods for consumption, and yield highly acceptable and diversified flavors. Depending on the salt concentration, salting directs the subsequent course of the fermentation, limiting the amount of pectinolytic and proteolytic hydrolysis that occurs, thereby controlling softening and preventing putrefaction. Lactic acid fermentations have other distinct advantages in that the foods become resistant to microbial spoilage and toxin development. Acid fermentations also modify the flavor of the original ingredients and often improve nutritive value.

Because canned or frozen foods are mostly unavailable or too expensive for hundreds of millions of the world's economically deprived and hungry people, acid fermentation combined with salting remains one of the most practical methods of preservation, often enhancing the organoleptic and nutritional qualities of fresh vegetables, cereal gruels, and milk-cereal mixtures.

SAUERKRAUT

Lactic acid fermentation of cabbage and other vegetables is a common way of preserving fresh vegetables in the western world, China, and Korea (where *kimchi* is a staple in the diet). It is a simple way of preserving food: the raw vegetable is sliced or shredded, and approximately 2 percent salt is added. The salt extracts liquid from the vegetable, serving as a substrate for the growth of lactic acid bacteria. Anaerobic conditions should be maintained, insofar as possible, to prevent the growth of microorganisms that might cause spoilage.

The sequence of organisms that develop in a typical sauerkraut fermentation is as follows: *Leuconostoc mesenteroides* initiates the growth in the shredded cabbage over a wide range of temperatures and salt concentrations. It produces carbon dioxide and lactic and acetic acids, which quickly lower the pH, thereby inhibiting development of undesirable microorganisms that might destroy crispness. The carbon dioxide produced replaces the air and facilitates the anaerobiosis required for the fermentation. The fermentation is completed in sequence by *Lactobacillus brevis* and *Lb. plantarum*. *Lb. plantarum* is responsible for the high acidity. If the fermentation temperature or salt concentration is high, *Pecicoccus cerevisiae* develops and contributes to acid production.

As would be expected, the rate of completion of the fermentation depends on the temperature and salt concentration. At 7.5°C fermentation is very slow: under these circumstances, *L. mesenteroides* grows slowly, attaining an acidity of 0.4 percent in about 10 days and an acidity of 0.8 to 0.9 percent in a month. Lactobacilli and pediococci cannot grow well at this temperature, and the fermentation may not be completed for 6 months. At 18°C a total acidity (as lactic acid) of 1.7 to 2.3 percent will be reached, with an acetic to lactic acid ratio of 1:4, in about 20 days. At 32°C a similar activity will be reached in 8 to 10 days, with most of the acid being lactic acid produced by the homofermentative bacteria *Lb. plantarum* and *P. cerevesiae*.

Increasing the salt concentration to 3.5 percent results in 90 percent inhibition of growth and acid production for both *L. mesenteroides* and *Lb. brevis*. The ratio of nonvolatile to volatile acid produced has a marked effect on flavor, *Lb. brevis* producing a harsh, vinegar-like flavor and *L. mesenteroides* a mild, pleasantly aromatic flavor. The homofermenters *Lb. plantarum* and *P. cerevesiae* yield unacceptable products.

KOREAN *KIMCHI*

Korean *kimchi* differs from sauerkraut in two respects: it has, optimally, much less acid and it is carbonated. Chinese cabbage and radish are the major substrates; garlic, green onion, ginger, leaf mustard, hot pepper, parsley, and carrot are minor ingredients.

Kimchi is available year-round, is served three times daily, and is a diet staple along with cooked rice and certain side dishes. It accounts for about an eighth of the total daily food intake of an adult. Its popularity is largely due to its carbonation derived from fermentation with natural microflora.

Salting of the cabbage can be done at 5 to 7 percent salinity for 12 hours or 15 percent salinity for 3 to 7 hours, followed by rinsing and draining. Optimum salt concentration during *kimchi* fermentation is approximately 3 percent. Lower temperatures (about 10°C) are preferred to temperatures above 20°C. Optimum acidity of *kimchi* is 0.4 to 0.8 percent lactic acid with a pH between 4.2 and 4.5; higher acidity makes it unacceptable. Organisms isolated from *kimchi* include *L. mesenteroides, S. faecilis, Lb. brevis, Lb. plantarum,* and *P. cerevesiae*.

PICKLED VEGETABLES

Pickling of cucumbers and other vegetables is widely practiced today. Although a variety of techniques are used, placing cucumbers in a 5 percent salt brine is a satisfactory method. The cucumbers absorb salt until there is an equilibrium between the salt in the cucumbers and the brine. Acidity reaches 0.6 to 1.0 (as lactic acid) with a pH of 3.4 to 3.6 in about 2 weeks, depending on the temperature.

In Malaysia the most common vegetables pickled are cucumbers, ginger, onion, leek, chili, bamboo shoots, and leafy tropical vegetables like mustard leaves. Young unripe fruits commonly pickled include mangoes, papaya, pineapple, and lime. In Egypt carrots, cucumbers, turnips, cauliflower, green and black olives, onions, and hot and sweet peppers are among the vegetables pickled. They are used as appetizers and served with practically every meal.

INDIAN *IDLI* AND *DOSA*

Indian *idli* is a small, white, acidic, leavened, steam-cooked cake made by lactic fermentation of a thick batter made from polished rice and dehulled black gram dhal, a pulse (*Phaseolus mungo*). The cakes are soft, moist, and spongy and have a pleasant sour flavor. *Dosa*, a closely related product, is made from the same ingredients, both finely ground. The batter is generally thinner, and *dosa* is fried like a pancake.

Idli fermentation is a process by which leavened bread-like products can be made from cereals other than wheat or rye and without yeast. The initial step in the fermentation is to wash both rice and black gram dhal. They are then soaked for 5 to 10 hours and drained. The coarsely ground rice and black gram are then combined with water and 1 percent salt to make a thick batter. The batter is fermented in a warm place (30 to 32°C) overnight, during which time acidification and leavening occur. The batter is then placed in small cups and steamed or fried as a pancake. The proportions of rice to black gram vary from 4:1 to 1:4, depending on the relative cost on the market.

Idli and *dosa* are both products of natural lactic acid fermentation. *L. mensenteroides* and *S. faecalis* develop during soaking, then continue to multiply following grinding. Each eventually reaches more than 1×10^9 cells per gram, 11 to 13 hours after formation of the batter. These two species predominate until 23 hours following batter formation. Practically all batters would be steamed by then. If a batter is further incubated, the lactobacilli and streptococci decrease in numbers and *P. cerevisiae* develops. *L. mesenteroides* is the microorganism essential for leavening of the batter and, along with *S. faecalis*, is also responsible for acid production. Both functions are essential for producing a satisfactory *idli*.

In *idli* made with a 1:1 ratio of black gram to rice, batter volume increased about 47 percent 12 to 15 hours after incubation at 30°C. The pH fell to 4.5 and total acidity rose to 2.8 percent (as lactic acid). Using a 1:2 ratio of black gram to rice, batter volume increased 113 percent and acidity rose to 2.2 percent in 20 hours at 29°C. Reducing sugars (as glucose) showed a steady decrease from 3.3 milligrams per gram of dry ingredients to 0.8 milligrams per gram in 20 hours, reflecting their utilization for acid and gas production. Soluble solids increased, whereas soluble nitrogen decreased. Flatulence-causing oligosaccharides, such as stachyose and raffinose, are completely hydrolyzed.

A 60 percent increase in methionine has been reported during fermentation. The increase would be of considerable nutritional importance if true, but the results conflict with earlier findings. Thiamine and riboflavin increases during fermentation and phytate phosphorous decreases have also been reported.

PHILIPPINE *PUTO*

Philippine *puto* is a leavened steamed rice cake made from year-old rice grains that are soaked, ground with water, and allowed to undergo a natural acid and gas fermentation. Part of the acid is neutralized with sodium hydroxide during the last stage of fermentation. *Puto* is closely related to Indian *idli*, except that it contains no legume.

SOURDOUGH BREADS AND RELATED FERMENTATIONS

There is a close relationship between yeasts and lactic acid bacteria in sourdough breads, soy sauce, *miso*, and *kefir*. Sourdough leaven contains both yeasts and lactobacilli. The method of preparing such leavens is ancient. Wheat, rye, or other cereal grain flour is mixed with water and incubated for a few days in a warm place. Initially, a wide range of microorganisms develop, but eventually the lactic acid bacteria predominate because of their acid production. Yeasts also can survive, because they tolerate acid well. More flour is added to make a dough. This dough is then subdivided and used to make a batch of bread, while the rest of the dough is kept for future bread making. Wherever sourdough leavens have been studied, the organisms found have been similar.

The essential microorganisms in sourdough are a *Lactobacillus* sp. and a yeast, *Torulopsis holmii*. *Saccharomyces inusitatus* also has been isolated and identified in sourdough leaven. The lactobacillus species has a preference for maltose and uses the maltose phosphorylase pathway to metabolize the sugar, whereas *T. holmii* grows on glucose but not on maltose, so that both develop in a dough where the amylases hydrolyze starch to maltose.

The basic biochemical changes that occur in sourdough bread fermentation are (1) acidification of the dough with lactic and acetic acids produced by the lactobacilli and (2) leavening of the dough with carbon dioxide produced by the yeast and the lactobacilli. Typical flavor and aroma development can be traced to biochemical activities of both lactobacilli and yeasts. The chewy characteristic of sourdough bread may be due to the production of bacterial polysaccharides by the lactobacilli.

NIGERIAN *OGI* (KENYAN *UJI*)

Nigerian *ogi* is a smooth-textured, sour porridge with a flavor resembling that of yogurt. It is made by lactic acid fermentation of corn, sorghum, or millet. Soybeans may be added to improve nutritive value. *Ogi* has a solids content of about 8 percent. The cooked gel-like porridge is known as "pap."

The first step in the fermentation is steeping of the cleaned grain for 1 to 3 days. During this time the desirable microorganisms develop and are selected. The grain is then ground with water and filtered to remove coarse particles. After steeping, the pH should be 4.3. Optimum pH for *ogi* is 3.6 to 3.7. The concentration of lactic acids may reach 0.65 percent and that of acetic acid 0.11 percent during fermentation. If the pH falls to 3.5, it is less acceptable.

Ogi is a naturally fermented product. A wide variety of molds, yeasts, and bacteria are present initially. *Lb. plantarum* appears to be the essential microorganism in the fermentation. Following depletion of the fermentable sugars, it is able to utilize dextrins from the corn. *Saccharomyces cerevisiae* and *Candida mycoderma* contribute to the pleasant flavor.

NIGERIAN *GARI*

Nigerian *gari* is a granular starchy food made from cassava (*Manihot utilissima* or *M. esculenta*) by lactic acid fermentation of the grated pulp, followed by dry-heat treatment to gelatinize and semidextrinize the starch, which is followed by drying. Cassava tubers are washed, peeled, and grated. An inoculum of 3-day-old cassava juice or fermented mash liquor is added. The pulp is placed in a cloth bag, excess water is squeezed out, and the pulp undergoes an anaerobic acid fermentation for 12 to 96 hours. Optimum temperature is 35°C. When the pH of the mash reaches 4.0, with about 0.85 percent total acid (as lactic acid), the *gari* has the desired sour flavor and a characteristic aroma. In village processes, further moisture may be removed, and the pulp is then toasted (semidextrinized) in shallow iron pots and dried to less than 20 percent moisture. Village-processed *gari* has a carbohydrate content of about 82 percent with 0.9 percent protein. Lactic, acetic, propionic, succinic, and pyruvic acids have been identified in *gari*, with aldehydes and esters providing the aroma.

For consumption the *gari* is added to boiling water, in which it increases in volume by 300 percent to yield a semisolid plastic dough. The stiff porridge is rolled into a ball (10 to 30 grams wet weight) with the fingers and dipped into stew.

PHILIPPINE *BALAO BALAO*

Balao balao is a lactic acid fermented rice-shrimp mixture, generally prepared by blending cooked rice, whole raw shrimp, and solar salt and then allowing the mixture to ferment for several days or weeks, depending on the salt content. The chitinous shell becomes soft, and when the fermented product is cooked, the whole shrimp can be eaten.

With a salt concentration of 3 percent added to the rice-shrimp mixture, the pH falls to an organoleptically desirable value of 4.08, with titratable acidity reaching 1.32 percent acid (as lactic acid) in 4 days.

Balao balao made with 3 percent salt is best in color, odor, flavor, texture, and general acceptability and is the least salty. *Balao balao* offers a basic method of preservation for cereal-shrimp-fish mixtures. When properly packed to exclude air, sufficient acid is produced to preserve the products without resorting to high-temperature cooking.

MEXICAN *PULQUE*

Pulque is a white, acidic, alcoholic beverage made by fermentation of juice of *Agave* species, mainly *A. atrovirens* or *A. americana*, the century plants. It has been a national Mexican drink since the time of the Aztecs. *Pulque* plays an important role in the nutrition of low-income people in the semiarid regions of Mexico. The essential microorganisms in the *pulque* fermentation are *Lb. plantarum*, a heterofermentative *Leuconostoc*, *Sac. cerevisiae*, and *Zymomonas mobilis*.

The heterofermentative *Leuconostoc* plays the essential role of producing dextrans, which contribute a characteristic viscosity to *pulque* and also increase the acidity of the agave juice very rapidly, inhibiting growth of other less desirable bacteria. *Lb. plantarum* contributes to the final acidity of *pulque*. *Sac. cerevisiae* appears to be a major producer of ethanol, but *Z. mobilis* is considered to be the most important ethanol producer in *pulque*. Under anaerobic conditions, *Zymomonas* transforms 45 percent of the glucose to ethanol and carbon dioxide. It also produces some acetic acid, acetylmethylcarbinol, and some slime gums, which may contribute to the viscous nature of traditional *pulque*.

Soluble solids in the fresh agave juice decrease from 25-30 percent to 6.0 percent in *pulque*. The pH falls from 7.4 to 3.5-4.0. Total acid increases from 0.03 percent to 0.4-0.7 percent (as lactic acid). Sucrose decreases from 18.6 percent to less than 1 percent. Ethanol increases from 0 percent to 4-6 percent (v/v). The B vitamins are present in nutritionally important quantities, with ranges reported as follows (in milligrams per 100 grams): thiamine, 5 to 29; niacin, 54 to 515; riboflavin, 18 to 33; pantothenic acid, 60 to 335; p-aminobenzoic acid, 10 to 12; pyridoxine, 14 to 23; and biotin, 9 to 32.

EGYPTIAN *KISHK*, GREEK *TRAHANAS*, AND TURKISH *TARHANAS*

Egyptian *kishk*, Greek *trahanas*, and Turkish *tarhanas* are mixtures of sheep's milk yogurts and parboiled wheat. Tomato, tomato paste, or onion are sometimes added. In all cases the milk or buttermilk undergoes a typical lactic acid fermentation in which the pH ranges from 3.5 to 3.8 and titratable acidity is 1.3 to 1.8 percent (as lactic acid). Proportions of wheat to yogurt range from 2:1 to 1:3. The wheat is parboiled at some stage in the process. In its simplest form the wheat is added directly to the yogurt and the mixture is boiled until the wheat has absorbed the free moisture. The mixture is cooled and formed into biscuits that are sun dried. If the wheat is ground prior to mixing with the yogurt, the fines are discarded because they harden the final product.

In Egypt the principal microorganisms reported in *kishk* are the heterofermentative *Lb. brevis* and the homofermentative *Lb. casei* and *Lb. plantarum*. In Cyprus sheep's milk yogurt contains principally *S. thermophilus* and *Lb. bulgaricus*. Dried *kishk* and *trahanas* are not hygroscopic and can be stored in open jars for several years without deterioration. They also are well balanced nutritionally.

OTHER FOODS

Lactic acid fermentation also plays an essential role in the production of Indonesian *tempe*, a vegetable (soybean) protein meat substitute the texture of which is provided by mycelium of *Rhizopus oligosporus*, which overgrows and knits the soaked, partially cooked cotyledons into compact cakes that can be sliced thinly and deep fried or cut into chunks and used in soups in place of meat. The essential part played by lactobacilli occurs during the initial soaking when the pH falls from about 6.5 to between 4.5 and 5.0. The lower pH facilitates growth of the mold and prevents development of undesirable bacteria that might spoil the *tempe*.

In Chinese soy sauce (Japanese *shoyu)* and Japanese *miso* and related meat-flavored, amino acid peptide sauces and pastes, the essential microorganism for amylolytic, proteolytic hydrolysis of the soybean-wheat or soybean-rice or barley substrates is *Aspergillus oryzae*. Following overgrowth of the substrate by the mold, the *koji* is subsequently allowed to ferment in approximately 19 percent salt brine for the sauces and 6 to 13 percent salt for the pastes. Lactobacilli grow and lower the pH to about 4.5, which then allows the osmophilic yeast *Sac. rouxii* to grow and produce some ethanol. The ethanol combines with organic acid in the substrate, producing esters that contribute to the agreeable flavor and aroma.

Given the fact that these acid fermentation techniques are simple, effective, and inexpensive, their application in developing countries should be encouraged.

6
Mixed-Culture Fermentations

Clifford W. Hesseltine

Mixed-culture fermentations are those in which the inoculum always consists of two or more organisms. Mixed cultures can consist of known species to the exclusion of all others, or they may be composed of mixtures of unknown species. The mixed cultures may be all of one microbial group—all bacteria—or they may consist of a mixture of organisms of fungi and bacteria or fungi and yeasts or other combinations in which the components are quite unrelated. All of these combinations are encountered in Oriental food fermentations.

The earliest studies of microorganisms were those made on mixed cultures by van Leeuwenhoek in 1684. Micheli, working with fungi in 1718, reported his observations on the germination of mold spores on cut surfaces of melons and quinces. In 1875 Brefeld obtained pure-culture of fungi, and in 1878 Koch obtained pure cultures of pathogenic bacteria. The objective of both Brefeld's and Koch's studies was to identify pathogenic microorganisms. They wanted to prove what organism was responsible for a particular disease. Thus, part of Koch's fame rests on his discovery of the cause of tuberculosis.

An early paper on mixed-culture food fermentation was an address by Macfadyen (1) at the Institute of Brewing, in London, in 1903 entitled, "The Symbiotic Fermentations," in which he referred to mixed-culture fermentations as "mixed infections." Probably this expression reflected his being a member of the Jenner Institute of Preventive Medicine. About half of his lecture was devoted to mixed-culture fermentations of the Orient. Among those described were Chinese yeast, *koji*, Tonkin yeast, and *ragi*.

Mixed cultures are the rule in nature; therefore, one would expect this condition to be the rule in fermented foods of relatively ancient origin. Soil, for example, is a mixed-organism environment with protozoa, bacteria, fungi, and algae growing in various numbers and kinds, depending on the nutrients available, the temperature, and the pH of the soil. Soil microorganisms relate to each other—some as parasites on others, some forming substances essential to others for growth, and some having no effect on each other.

ADVANTAGES

Mixed-culture fermentations offer a number of advantages over conventional single-culture fermentations:

- Product yield may be higher. Yogurt is made by the fermentation of milk with *Streptococcus thermophilus* and *Lactobacillus bulgaricus*. Driessen (2) demonstrated that when these species were grown separately, 24 mmol and 20 mmol, respectively, of acid were produced; together, with the same amount of inoculum, a yield of 74 mmol was obtained. The number of *S. thermophilus* cells increased from 500×10^6 per milliliter to 880×10^6 per milliliter with *L. bulgaricus*.
- The growth rate may be higher. In a mixed culture one microorganism may produce needed growth factors or essential growth compounds such as carbon or nitrogen sources beneficial to a second microorganism. It may alter the pH of the medium, thereby improving the activity of one or more enzymes. Even the temperature may be elevated and promote growth of a second microbe.
- Mixed cultures are able to bring about multistep transformations that would be impossible for a single microorganism. Examples are the *miso* and *shoyu* fermentations in which *Aspergillus oryzae* strains are used to make *koji*. Koji produces amylases and proteases, which break down the starch in rice and proteins in soybeans. In the *miso* and *shoyu* fermentations, these compounds are then acted on by lactic acid bacteria and yeast to produce flavor compounds and alcohol.
- In some mixed cultures a remarkably stable association of microorganisms may occur. Even when a mixture of cultures is prepared by untrained individuals working under unsanitary conditions, such as in *ragi*, mixtures of the same fungi, yeasts, and bacteria remain together even after years of subculture. Probably the steps in making the starter were established by trial and error, and the process conditions were such that this mixture could compete against all contaminants.
- Compounds made by a mixture of microorganisms often complement each other and work to the exclusion of unwanted microorganisms. For example, in some food fermentations yeast will produce alcohol and lactic acid bacteria will produce lactic acid and other organic acids and change the environment from aerobic to anaerobic. Inhibiting compounds are thus formed, the pH is lowered, and anaerobic conditions are developed that exclude most undesirable molds and bacteria.

• Mixed cultures permit better utilization of the substrate. The substrate for fermented food is always a complex mixture of carbohydrates, proteins, and fats. Mixed cultures possess a wider range of enzymes and are able to attack a greater variety of compounds. Likewise, with proper strain selection they are better able to change or destroy toxic or noxious compounds that may be in the fermentation substrate.

• Mixed cultures can be maintained indefinitely by unskilled people with a minimum of training. If the environmental conditions can be maintained (i.e., temperature, mass of fermenting substrate, length of fermentation, and kind of substrate), it is easy to maintain a mixed-culture inoculum indefinitely and to carry out repeated successful fermentations.

• Mixed cultures offer more protection against contamination. In mixed-culture fermentations phage infections are reduced. In pure-culture commercial fermentations involving bacteria and actinomycetes, invariably an epidemic of phage infections occurs, and the infection can completely shut down production. Since mixed cultures have a wider genetic base of resistance to phage, failures do not occur, often because if one strain is wiped out, a second or third phage-resistant strain in the inoculum will take over and continue the fermentation. In such processes, especially with a heavy inoculum of selected strains, contamination does not occur even when the fermentations are carried out in open pans or tanks.

• Mixed-culture fermentations enable the utilization of cheap and impure substrates. In any practical fermentation the cheapest substrate is always used, and this will often be a mixture of several materials. For example, in the processing of biomass, a mixed culture is desirable that attacks not only the cellulose but also starch and sugar. Cellulolytic fungi along with starch- and sugar-utilizing yeasts would give a more efficient process, producing more product in a shorter time.

• Mixed cultures can provide necessary nutrients for optimal performance. Many microorganisms, such as the cheese bacteria, which might be suitable for production of a fermentation product, require growth factors to achieve optimum growth rates. To add the proper vitamins to production adds complications and expense to the process. Thus, the addition of a symbiotic species that supplies the growth factors is a definite advantage.

DISADVANTAGES

Mixed-culture fermentations also have some disadvantages.

• Scientific study of mixed cultures is difficult. Obviously, it is more difficult to study the fermentation if more than one microorganism is involved. That is why most biochemical studies are conducted as single-culture fermentations because one variable is eliminated.

- Defining the product and the microorganisms employed becomes more involved in patent and regulatory procedures.
- Contamination of the fermentation is more difficult to detect and control.
- When two or three pure cultures are mixed together, it requires more time and space to produce several sets of inocula rather than just one.
- One of the worst problems in mixed-culture fermentation is the control of the optimum balance among the microorganisms involved. This can, however, be overcome if the behavior of the microorganisms is understood and this information is applied to their control.

The balance of organisms brings up the problem of the storage and maintenance of the cultures. Lyophilization presents difficulties because in the freeze-drying process the killing of different strains' cells will be unequal. It is also difficult, if not impossible, to grow a mixed culture from liquid medium in contrast to typical fermentations on solid mediums, without the culture undergoing radical shifts in population numbers. According to Harrison (3), the best way to preserve mixed cultures is to store the whole liquid culture in liquid nitrogen below −80°C. The culture, when removed from the frozen state, should be started in a small amount of the production medium and checked for the desired fermentation product and the normal fermentation time. Subcultures of this initial fermentation, if it is satisfactory, may then be used to start production fermentations.

FUTURE

Mixed-culture fermentations will continue to be used in traditional processes such as soybean and dairy fermentations. As noted above, the extensive uses of mixed-culture fermentations for dairy and meat products are well known as to the type of cultures used and the fermentation process. However, there are a large number of food fermentations based on plant substrates such as rice, wheat, corn, soybeans, and peanuts in which mixed cultures of microorganisms are used and will continue to be used

One example of the complex sequential interaction of two fermentations, and which employs fungi, yeast, and bacteria, is the manufacture of *miso*. This Oriental food fermentation product is based on the fermentation of soybeans, rice, and salt to make a paste-like fermented food. *Miso* is used as a flavoring agent and as a base for *miso* soup. There are many types of *miso*, ranging from a yellow sweet *miso* (prepared by a quick fermentation) to a dark, highly flavored *miso*. The type depends on the amount of salt, the ratio of cereals to soybeans, and the duration of the fermentation.

The *miso* fermentation begins with the molding of sterile, moist, cooked rice that is inoculated with dry spores of *Aspergillus oryzae* and *A. soyae*. The inoculum consists of several mold strains combined, with each strain producing a desired enzyme(s). The molded rice is called *koji* and is made to produce enzymes to act on the soybean proteins, fats, and carbohydrates in the subsequent fermentation.

After the rice is thoroughly molded, which is accomplished by breaking the *koji* and mixing, the *koji* is harvested before mold sporulation starts, usually in 1 or 2 days. The *koji* is mixed with salt and soaked and steamed soybeans. This mixture is inoculated with a new set of microorganisms, and the four ingredients are now mashed and mixed. After the production of *koji* with molds, the paste is placed in large concrete or wooden tanks for the second fermentation. The inoculum consists of osmophilic yeasts *Saccharomyces rouxii* and *Candida versatilis* and one or more strains of lactic acid bacteria, typically *Pediococcus pentosaceus* and *P. halophilus* (4). Conditions in the fermentation tanks are anaerobic or nearly so, with the temperature maintained at 30°C. The fermentation is allowed to proceed for varying lengths of time, depending on the type of *miso* desired, but it is typically 1 to 3 months. The fermenting mash is usually mixed several times, and liquid forms on the top of the fermenting mash.

The initial inoculum is about 10^5 microorganisms per gram. Typically, 3,300 kg of *miso* with a moisture level of 48 percent is obtained when 1,000 kg of soybeans, 600 kg of rice, and 430 kg of salt are used. When the second fermentation is completed, aging is allowed to take place. A number of other mixed-culture fermentations are similar to the *miso* process, including *shoyu* (soy sauce) and *sake* (rice wine).

A legitimate question can be asked as to the future prospects for the use of mixed cultures in food fermentations. What will be the effect of genetic engineering on the use of mixed cultures? Would engineered organisms be able to compete in mixed culture? Many laboratories are busy introducing new desirable genetic material into a second organism. The characteristics being transferred may come from such diverse organisms as mammals and bacteria and may be transferred from animals to bacteria. In general, the objective of this work involves introduction of one desirable character, not a number. For instance, strains of *Escherichia coli* have been engineered to produce insulin. However, I suspect that it may be a long time, if ever, before a single organism can produce the multitude of flavors found in foods such as cheeses, soy sauce, *miso*, and other fermented foods used primarily as condiments. The reason for this is the fact that a flavoring agent such as *shoyu* contains literally hundreds of compounds produced by the microorganisms, products from the action of enzymes on the substrate, and compounds formed by the nonenzymatic interactions of the products with the original substrate compounds.

To put such a combination of genes for all these flavors into one microorganism would, at present, be almost impossible. Second, the cost of producing the food, which is relatively inexpensive as now produced, would become economically prohibitive. The use of mixed cultures in making fermented foods from milk, meat, cereals, and legumes will continue to be the direction in the future.

Harrison (3), in his summary of the future prospects of mixed-culture fermentations, very succinctly concluded as follows:

> No claim for novelty can be made for mixed cultures: They form the basis of the most ancient fermentation processes. With the exploitation of monocultures having been pushed to its limits it is perhaps time to reappraise the potential of mixed culture systems. They provide a means of combining the genetic properties of species without the expense and dangers inherent in genetic engineering which, in general terms, aims at the same effect.

REFERENCES

1. Macfadyen, A. 1903. The symbiotic fermentations. Journal of the Federal Institutes of Brewing 9:2-15.

2. Driessen, F. M. 1981. Protocooperation of yogurt bacteria in continuous culture. Pp. 99–120 in: Mixed Culture Fermentations. M. E. Bushell and J. H. Slater, Eds. London: Academic Press.

3. Harrison, D. E. F. 1978. Mixed cultures in industrial fermentation processes. Advances in Applied Microbiology 24:129-164.

4. Hesseltine, C. W. 1983. Microbiology of oriental fermented foods. Annual Reviews of Microbiology 37:575-601.

III. MILK DERIVATIVES

7

Fermented Milks—Past, Present, and Future

M. Kroger, J. A. Kurmann, and J. L. Rasic

Milk is the most important foodstuff for a mammal and has always been the first food of the newborn. One could argue that the deliberate souring or fermentation of milk was one of the key achievements that nurtured mankind to grow and develop into a productive and preeminent species. Had fermented milk been considered spoiled and inedible and thus not have entered the human diet in the thousands of years to come, human development would have taken an entirely different course. Although there is no perfect food, milk is the most nearly perfect food known.

At some stage in the course of human evolution it was recognized that the milk of other mammals was equally satisfying in meeting physiological demands for moisture, energy, and nutrients. Milk from eight species of domesticated mammals (cow, buffalo, sheep, goat, horse, camel, yak, and zebu) has been used to make traditional fermented milk products throughout the world.

From a biological standpoint, fermented milks are characterized by the accumulation of microbial metabolic products. It was realized very early that such microbial metabolites as lactic acid, ethyl alcohol, and dozens of other chemicals collectively called flavor substances, were not altogether unpleasant and even contributed to overall preservative action.

CLASSIFICATIONS

Despite the long historical record and worldwide distribution of fermented milks, few people know more than five or 10 of the several hundred specific products that could be described. Even current food science and dairy technology textbooks fail to do the subject justice.

For example, the latest (fourth) edition of *Food Microbiology* (1) covers fermented dairy products in only two pages. The textbook used in the Pennsylvania State University dairy technology course is *The Science of Providing Milk for Man* (2). Cultured and acidified milk products occupy 10 pages, and cultured buttermilk, sour cream, yogurt, acidophilus milk, and *ymer* and lactofil are given only subchapter status. *Koumiss* and *kefir* are merely mentioned as being popular in Eastern Europe. *Cheese and Fermented Milk Foods* (3) is somewhat more comprehensive, but it deals mainly with practical concerns and primarily with cheese.

By far the best compilations on fermented milks have been and are being published as documents of the International Dairy Federation (4,5). One chapter of the latter lists some 80 fermented milks, including both traditional and nontraditional products. A soon-to-be-published encyclopedia of fermented fresh milk products (6) describes some 200 traditional fermented milks and several hundred nontraditional ones.

Traditional and Nontraditional

The most fundamental division of fermented milk products is into traditional and nontraditional types. Traditional fermented milk products have a long history and are known and made all over the world whenever milk animals were kept. Their production was a crude art. It was not until the days of Pasteur—about 100 years ago—that the microbiology underlying fermentations was revealed. In contrast, nontraditional fermented milk products are recently developed. They are based on known scientific principles; their microbial cultures are known; and their quality can be optimized. This is not the case with traditional products made with ill-defined, empirical cultures where you have to take what you get out of the fermentation. Yogurt is both a traditional and a nontraditional product—the latter being represented by ever-changing varieties.

Medium and Procedure

Classification by technology differentiates between fermented milks and fermented products not based directly on milk. It is obvious that products other than fresh milk can serve as the fermentation medium or substrate, such as cream, whey, buttermilk, and dry milk solids. It is also possible to further manipulate or change the curd recovered after coagulation.

Further Processing

Neither law nor taboo forbids experimentation with fermented milks. Numerous products are known that are mixtures of milk and other foodstuffs and that have been subjected to fermentation. These include fermented milk-vegetable products, fermented milk-meat extract mixtures, and fermented milk-fishmeal hydrolyzate mixtures. Consequently, we find societies that have utilized specific plants, meat extracts, or fishmeal hydrolyzates to enhance their nutritional status and the flavor and variety of their cuisine.

Pharmaceutical preparations are unique in that they emphasize microorganisms only instead of milk nutrients or product flavor. The subject of probiotics (a word coined in 1974) will undoubtedly emerge as a major field of study. We see it in animal science now where some work is being done to get specific bacteria implanted or colonized in the gastrointestinal tract of animals, obviously in the interest of animal health and improvement of farm animal food production. So-called health food stores make available preparations that provide people with specific doses of bacteria, such as *Lactobacillus acidophilus*, commonly found in some fermented milk products. The subjects of health and probiotics, as well as myth and faddism, are beyond the scope of this paper.

End Uses

Traditionally, fermented milk products have been consumed as beverages, as meal components, or as ingredients in cookery. As social patterns have changed, however, meal eaters have become snackers and grazers. Furthermore, food technologists and food innovators have created a multitude of new products for the shelves of modern supermarkets. Most of the developments have been in the dessert and confectionery category.

Microbial Actions

Homemade fermented milk products, especially in nomadic or village environments, are still occasionally made by spontaneous fermentation, but most likely they are made by the use of an empirical culture. In other words, the inoculum is obtained from a previous production and its microbial identity is unknown.

The bacteria utilized are either mesophiles or thermophiles, terms indicating optimum bacterial growth temperatures, roughly 70° and 100°F (22° and 38°C), respectively. More specific and important is the bacterial species present. A fermented milk is mainly characterized by its sensory properties, and the sensory properties, such as taste, odor, and viscosity, are the direct results of specific bacterial action. The current names of microorganisms recognized in fermented milks are listed in Table 1.

TABLE 1 Current Names of Microorganisms in Fermented Milks

Current Name	Number of Former Designations and Synonyms
Genus Lactobacillus	
L. delbrueckii	8
L. delbrueckii subsp. lactis	10
L. delbrueckii subsp. bulgaricus	8
L. acidophilus	2
L. helveticus	7
L. casei	6
L. brevis	29
L. fermentum	10
L. kefir	4
Genus Leuconostoc	
L. mesenteroides	2
L. mesenteroides subsp. dextranicum	6
L. mesenteroides subsp. cremoris	3
L. lactis	1
Genus Pediococcus	
P. pentosaceous	3
P. acidilactici	1
Genus Propionibacterium	
P. freudenreichii subsp. shermanii	
P. freudenreichii subsp. freudenreichii	
Genus Streptococcus	
S. lactis	1
S. lactis subsp. diacetylactis	1
S. lactis subsp. cremoris	3
S. thermophilus	1
Genus Bifidobacterium	
B. bifidum	13
B. longum[1]	1
B. infantis	3
B. breve	2
Genus Acetobacter	
A. aceti	11
Yeasts	
Torulaspora delbrueckii	1
Kluyveromyces marxianus subsp. marxianus	3
Kluyveromyces marxianus subsp. bulgaricus	1
Candida kefyr	3
Saccharomyces cerevisiae	0

[1]In an earlier edition of Bergey's Manual, *B. longum* was listed as having two subspecies: *B. longum* subsp. *longum* and *B. longum* subsp. *animalis*. The latter was translocated in the new Bergey's into two species: *B. animalis* and *B. pseudolongum*.

With regard to bacterial species, a number of products have evolved that are now characterized by the presence of specific organisms. Modern yogurt is now defined by the regulations of many governments to be made from and to contain only *Lactobacillus bulgaricus* and *Streptococcus thermophilus*. But there are no hard-and-fast rules, and, theoretically, any combination of organisms could be utilized to make a fermented milk product. The ultimate test is palatability. Frankly, there is still much confusion over the microbial identity of most of the known traditional fermented milk products in the world. Some have never been studied in depth. Some are very variable from batch to batch. Only yogurt has been given a proper definition by regulatory authorities in some countries. All other products are only loosely defined.

RESEARCH

Milk has always turned sour, but at some point in human history artisans deliberately caused milk to coagulate. However, the scientific principles behind the phenomenon of milk fermentation have remained unrevealed until recent decades.

We had to wait for the pioneers in microbiology to lead the way. Louis Pasteur (1822–1895) studied alcohol fermentation; Heinrich Anton DeBary (1831–1888) studied the infection of plants by fungi; and Robert Koch (1843–1910) studied human disease caused by bacteria. It was Elie Metchnikoff (1845–1916) who, while working at the Pasteur Institute in Paris, moved milk fermentations and the unheard-of subject of probiotics into the limelight. In 1908 he shared the Nobel Prize in Physiology and Medicine. Metchnikoff developed a theory that lactic acid bacteria in the digestive tract could, by preventing putrefaction, prolong life. His book, *The Prolongation of Life* (7), was translated into English in 1907 (reviewed in *Harper's Weekly*, February 8, 1908) and received much exposure worldwide. In a way it made Metchnikoff the godfather to everyone who, to this day, believes in the therapeutic value of fermented milk.

World War I put a damper on this type of human diet/health preoccupation. In the United States, it was 1921 before an American figure emerged who should be given much more credit, Leo Frederick Rettger. Rettger was a professor of bacteriology at Yale for most of his career. Two of his publications are *A Treatise on the Transformation of the Intestinal Flora with Special Reference to the Implantation of Bacillus Acidophilus* (8) and *Lactobacillus Acidophilus and Its Therapeutic Application* (9).

On the practical front at that time, A. D. Burke, head of the Dairy

Department of Alabama Polytechnic Institute, published *Practical Manufacture of Cultured Milks and Kindred Products* (10). Burke's book is, according to the subtitle, "a complete and practical treatise on the manufacture of commercial cultured buttermilks of all types—lactic, bulgarian, acidophilus, kefir, kumiss, yogurt." It is also a practical treatise on commercial casein, cottage cheese, cream cheese, and commercial sour cream, with information on dried, condensed, and fruit-flavored buttermilk.

Then came World War II, and until about 1950 very little research and development was seen on fermented milks. Since then increasing attention has been paid to fermented milk products worldwide. The American Cultured Dairy Products Institute was created in the United States in 1965. Several good books have been published, and scientific publications on the subject are proliferating. Manufacturers, researchers, and the public are experimenting with cultured dairy products in North America—and not only with yogurt but with other products as well. *Kefir* has been available in Los Angeles for more than a decade. In 1985 a New Jersey corporation began producing *kefir* for the East Coast, and in 1987 several major grocery chains began selling *leben*.

The future of fermented milk in North America and elsewhere will undoubtedly be exciting and complex.

REFERENCES

1. Frazier, W. C., and D. C. Westhoff. 1987. *Food Microbiology*. 4th ed. New York: McGraw-Hill Book Co.

2. Campbell, J. R., and R. T. Marshall. 1975. *The Science of Providing Milk for Man*. New York: McGraw-Hill Book Co.

3. Kosikowski, F. V. 1977. *Cheese and Fermented Milk Foods*, 2nd ed. Ann Arbor, Mich.: Edwards Brothers, Inc.

4. International Dairy Federation. 1984. Fermented Milks. Document 179, International Dairy Federation, Brussels, Belgium.

5. International Dairy Federation. 1989. Monograph on Fermented Milks: Science and Technology. International Dairy Federation, Brussels, Belgium.

6. Kurmann, J. A., J. L. Rasic, and M. Kroger. 1990. *Encyclopedia of Fermented Fresh Milk Products*. New York: Van Nostrand Reinhold.

7. Metchnikoff, E. 1906. *The Prolongation of Life*. New York: G. P. Putnam and Sons.

8. Rettger, L. F., and H. A. Cheplin. 1921. *A Treatise on the Transformation of the Intestinal Flora with Special Reference to the Implantation of Bacillus Acidophilus*. New Haven, Conn.: Yale University Press.

9. Rettger, L. F., M. N. Levy, L. Weinstein, and J. E. Weiss. 1935. *Lactobacillus Acidophilus and Its Therapeutic Application*. New Haven, Conn.: Yale University Press.

10. Burke, A. D. 1938. *Practical Manufacture of Cultured Milks and Kindred Products*. Milwaukee, Wis.: The Olsen Publishing Co.

8

Lactobacillus GG Fermented Whey and Human Health

Seppo Salminen and Kari Salminen

Traditionally, whey has been a troublesome waste product at cheese factories. New uses have now been developed for cheese whey to utilize the whey nutrients, including protein and carbohydrates.

Fermented milk products have been reported to have an important role in the treatment of infant diarrhea in malnourished children (1,2). More recently, Isolauri and co-workers (3) have shown in a double-blind controlled trial that *Lactobacillus* GG bacteria promote recovery from acute diarrhea in children. These results suggest that whey-based products may be used in this application.

A process for manufacturing a fermented flavored whey drink has been developed that combines the nutritional properties of whey and the health benefits of *Lactobacillus* strain GG. The objective has been to improve the utilization of whey through use of a scientifically selected *Lactobacillus* strain with proven health benefits. For this purpose, demineralized lactose-hydrolyzed whey concentrate has been fermented with *Lactobacillus* GG. Whey and lactic acid bacteria have thus been combined to provide a wholesome and nutritious beverage.

WHEY HYDROLYSIS PROCESS

Important steps in whey processing are the hydrolysis of lactose and demineralization to remove excess salt. A continuous whey hydrolysis process has been developed using immobilized ß-galactosidase enzyme. This process is more economical than batch hydrolysis. Lactose hydrolysis is important for lactose-intolerant populations and for malnourished children. Malnourished children may experience worsening of acute diarrhea when lactose is given during treatment (1). Salt removal can be completed using an ion exchange process. After concentration to 60 percent dry matter, a hydrolyzed demineralized whey syrup is obtained that has a good shelf life and a pleasant rich taste.

LACTOBACILLUS GG

Lactobacillus casei strain GG (*Lactobacillus* GG) is a new *Lactobacillus* strain that is of human origin and has been shown to colonize the intestinal tract (4). This strain was originally isolated from a healthy human volunteer based on its ability to tolerate acid and bile, to produce an antimicrobial substance, and to adhere to human intestinal cells (5,6). It is among the first strains with clinically proven health benefits in various intestinal disorders in adults, children, and infants. The most important evidence of its health benefits comes from studies of infant diarrhea. Isolauri and co-workers (3) published the first study on infant rotavirus diarrhea in which the duration of diarrhea was reduced by 50 percent through the use of either freeze-dried *Lactobacillus* GG or *Lactobacillus* GG fermented milk products.

LACTOBACILLUS GG FERMENTED WHEY DRINK

A new fermented flavored whey drink has been manufactured from demineralized lactose-hydrolyzed whey concentrate using *Lactobacillus* GG. It is a low-lactose product that contains no fat and is lightly sweetened with fructose. It has special sensory characteristics—smooth texture, mild acidity, and the rich taste from whey. Fruit juices or fruit flavoring have been used to modify the flavor to appeal to different people.

Fermentation of whey may also influence lactose content when suitable bacteria are used. Additionally, whey proteins may undergo slight changes to ease their digestibility. The end product may offer alternatives for people not currently attracted to fermented milks.

CONCLUSIONS

This development in whey processing offers new alternatives for utilizing cheese by-products and applies new technologies to nutritionally important products. Combining whey processing with lactobacilli that have been obtained using new selection methods may prove to be beneficial to human health in many intestinal imbalances. It may also offer possibilities in utilizing new technologies in food production in different cultures and in providing nutritionally attractive foods from low-value by-products.

REFERENCES

1. Bhan, M. K., S. Sazawal, S. Bhatnagar, B. L. Jailkhani, and N. Arora. 1989. Efficacy of yoghurt in comparison to milk in malnourished children with acute diarrhea. Pp. 229–232 in: Les laits fermentés: Actualité de la recherche. U.K. John Libbey Eurotext Ltd.

2. Boudraa, G., M. Touhami, P. Pochard, R. Soltana, J. Y. Mary, and J. F. Desjeux. 1990. Effect of feeding yogurt versus milk in children with persistent diarrhea. Journal of Pediatric Gastroenterology and Nutrition 11:409–512.

3. Isolauri, E., M. Juntunen, T. Rautanen, P. Sillanaukee, and T. Koivula. 1991. A human *Lactobacillus* strain (*Lactobacillus casei* strain GG) promotes recovery from acute diarrhea in children. Pediatrics 88:90–97.

4. Saxelin, M., S. Elo, S. Salminen, and H. Vapaatalo. 1990. Dose response colonisation of faeces after oral administration of *Lactobacillus casei* strain GG. Microbial Ecology 4:209–214.

5. Silva, M., N. Jacobus, C. F. Deneke, and S. Gorbach. 1987. Antimicrobial substance from a human *Lactobacillus* strain. Antimicrobial Agents and Chemotherapy 31:1231–1233.

6. Elo, S., M. Saxelin, and S. Salminen. 1991. Attachment of *Lactobacillus casei* strain GG to human colon carcinoma cell line Caco-2: Comparison with other dairy strains. Letters on Applied Microbiology 13:154–156.

7. Siitonen, S., H. Vapaatalo, S. Salminen, A. Gordin, M. Saxelin, R. Wikberg, and A.M. Kirkkola. 1990. Effect of *Lactobacillus* GG yoghurt in prevention of antibiotic associated diarrhoea. Annales Medicinae 22:57–60.

9
The Microbiology of Ethiopian *Ayib*

Mogessie Ashenafi

In Ethiopia, smallholder milk processing is based on sour milk resulting from high ambient temperatures, while meeting consumers' preferences and improving keeping quality (1). *Ayib*, a traditional Ethiopian cottage cheese, is a popular milk product consumed by the various ethnic groups of the country. It is made from sour milk after the butter is removed by churning. Traditional *ayib* making has been described by O'Mahony (1). Milk for churning is accumulated in a clay pot over several days. This is kept in a warm place (about 30°C) for 24 to 48 hours to sour spontaneously. Churning of the sour milk is carried out by slowly shaking the contents of the pot until the butter is separated. The butter is then removed from the churn and kneaded with water. The casein and some of the unrecovered fat in skim milk can be heat precipitated to a cottage cheese known as *ayib*. The defatted milk is heated to about 50°C until a distinct curd forms. It is then allowed to cool gradually, and the curd is ladled out or filtered through a muslin cloth. Temperature can be varied between 40° and 70°C without markedly affecting product composition and yield. Heat treatment does not appear to affect yield but gives the product a cooked flavor.

AYIB CHARACTERISTICS

Ayib comprises about 79 percent water, 15 percent protein, 2 percent fat, 1 percent ash, and 3 percent soluble milk constituents. The yield should be about 1 kilograms of *ayib* from 8 liters of milk (1).

The safety of cheese with respect to food-borne diseases is of great concern around the world. This is especially true in developing countries, where production of milk and various dairy products often takes place under unsanitary conditions. Since there was no published information on the microbiology of milk and milk products in Ethiopia, a study was carried out in our laboratory to evaluate the microbiological quality of *ayib* as available to the consumer (2). One hundred samples of *ayib* were purchased at the Awassa market over 10 weeks. Since Awassa is an open-air market, *ayib* was generally handled at ambient temperatures (about 25° to 27°C during the study period). Samples were microbiologically analyzed within two hours of purchase.

Standard microbiological procedures were followed to determine the counts of aerobic mesophilic microorganisms, psychrotrophs, yeasts and molds, coliforms, bacterial spores, enterococci, *Bacillus cereus, Listeria monocytogenes*, staphylococci, and lactic acid bacteria. The pH of the samples was also measured.

Ayib samples showed high numbers of mesophilic bacteria, enterococci, and yeasts (Table 1). More than 90 percent of the samples had aerobic mesophilic counts of 10^8 cfu/g (colony forming units) or higher; more than 75 percent of the samples had yeast counts of 10^7 cfu/g or higher, and over 85 percent contained enterococci in numbers of 10^7 cfu/g or higher. The majority of the samples had mold and lactic acid bacteria counts of 10^5 cfu/g or higher, spore-formers of about 10^4, and psychrotrophs of about 10^6 cfu/g (Table 1).

Over 32 percent had coliform counts of more than 10^2/g, and about 27 percent contained fecal coliform loads of more than 10^2/g. *Listeria* spp. were not detected from the samples. *B. cereus* and *S. aureus* were isolated in 63 percent and 23 percent of the samples, respectively, but at very low numbers (10^2 to 10^3 cfu/g). About 40 percent of the *ayib* samples had pH values of less than 3.7, and 60 percent had values of 3.7 to 4.6.

Most of the production of milk and various milk products in Ethiopia is generally a household process that usually takes place under unsanitary conditions. However, despite its high moisture content, the low pH of *ayib* may prevent the further proliferation of various microorganisms. Yeasts, which can grow at lower pH values, may affect the flavor and keeping quality of *ayib*. In another study (3), proteolytic yeasts made up 47 percent of the total yeast isolates and all isolates showed lipolytic activities. Since traditional *ayib* making involves removal of fat from the sour milk, *ayib* contains only about

TABLE 1 Frequency Distribution (Percent) of Aerobic Mesophilic Organisms, Yeasts, and Enterococci in *Ayib* Samples

	cfu/g[a]			
Microorganisms	$<10^7$	10^7	10^8	$>10^8$
Aerobic mesophils	—	8	65	27
Yeasts	25	70	5	—
Enterococci	13	45	37	5

[a]Colony forming units

1 percent fat, and thus the lipolytic isolates may not play an important role in affecting the flavor or keeping quality of *ayib*. Although proteolytic yeasts are important in cheese types that require ripening, their presence in a fresh product such as *ayib* is undesirable.

The findings in the previous studies indicated that *ayib* purchased from local markets was highly contaminated with various microorganisms. It was not known, however, whether these microorganisms were survivors of the heat treatment process or were postheating contaminants.

MODIFIED PROCESS

Another study was therefore conducted to determine the effect of cooking temperatures used in various parts of Ethiopia on the microbiological quality of the finished product and to recommend cooking temperatures that can decrease or destroy most microorganisms (4). *Ayib* was made in the laboratory using traditional methods. Pooled raw milk was allowed to sour naturally at room temperature. After removal of the fat by churning, the casein in the sour skimmed milk was heat precipitated at 40°, 50°, 60°, and 70°C in a water bath, and the curd was recovered by filtering through sterile cheese cloth.

Microbial analysis of raw milk, sour milk, and *ayib* indicated that heat treatment of the curd was effective at higher temperatures (Table 2). At these temperatures the time required for casein precipitation was also low. Heating the curd at 70°C for 55 minutes at pH 4 destroyed most of the microorganisms. The low pH also inhibited the proliferation of most surviving microorganisms. The high degree of contamination of market *ayib* could be due to either low curd cooking temperatures or addition of various plant materials to the finished product to give it desirable flavor, the packaging of *ayib* with Musa leaves, or other unhygienic handling practices. Thus, heat treatment of curd at 70°C and an appropriate handling of the product could result in a less contaminated and safer *ayib*.

TABLE 2 Frequency Distribution (Percent) of Lactic Acid Bacteria, Bacterial Spores, Molds, and Psychrotrophs in *Ayib* Samples

	[a]cfu/g			
Microorganisms	10^3	10^4	10^5	10^6
Lactic acid bacteria	14	33	13	40
Bacterial spores	4	68	24	4
Molds	12	25	35	28
Psychrotrophs	—	13	20	67

[a]Colony forming units

REFERENCES

1. O'Mahony, F. 1988. Rural Dairy Technology Experiences in Ethiopia. ILCA Manual No. 4, International Livestock Center for Africa, Addis Ababa.

2. Ashenafi, M. 1990. Microbiological quality of *ayib*, a traditional Ethiopian cottage cheese. International Journal of Food Microbiology 10:263-268.

3. Ashenafi, M. 1989. Proteolytic, lipolytic and fermentative properties of yeasts isolate from *ayib*, a traditional Ethiopian cottage cheese. SINET: Ethiopian Journal of Science 12:131-139.

4. Ashenafi, M. 1990. Effect of curd-cooking temperatures on the microbiological quality of *ayib*, a traditional Ethiopian cottage cheese. World Journal of Microbiology and Biotechnology 6:159-162.

10

Moroccan Traditional Fermented Dairy Products

Abed Hamama

In Morocco 20 to 30 percent of all milk produced is still processed by private individuals. These dairy shops and farmers manufacture traditional Moroccan dairy products such as *lben* and *raib* (fermented milks), *zabda* (farm butter), and *jben* (fresh cheese). These products are made from raw milk, and their physical properties are similar to those of commercially produced buttermilk, yogurt, butter, and fresh cheese. Although they are usually made from cow's milk, milk from sheep, goats, and camels also can be used. These products are very popular in Morocco mainly because of their refreshing qualities.

Basically, all these traditional dairy products are prepared by simply allowing the raw milk to ferment spontaneously at room temperature (15° to 25°C) for 1 to 3 days depending on the season. The coagulated milk is called *raib*. It can be consumed as such or churned in a clay jar to separate the liquid phase (*lben*) from fat (*zabda*). *Jben* is prepared by placing the coagulated milk in a cloth at room temperature and draining the whey. Salt is added to *jben* made in northern Morocco.

COMPOSITION AND MICROBIOLOGICAL CHARACTERISTICS

The Moroccan traditional fermented dairy products have been investigated (1-4) for their composition (Table 1) and their microbiology (Table 2). Data in these tables are average results only. In fact, a high level of variability for all the parameters was seen among samples of the same product. This heterogeneity is a consequence of the lack of standard procedures for preparation of these products.

Despite the acidic nature of these products (pH 4.0 to 4.5), they showed high counts of indicator microorganisms (e.g., coliforms, enterococci). This probably reflects poor hygienic conditions in the

TABLE 1 Average Physical-Chemical Composition of Moroccan Traditional Dairy Products

Composition	*Lben*(4)	*Jben*(3)	*Raib*(3)	*Zabda*(1)
pH	4.25	4.1	4.2	4.5
% lactic acid	0.81	1.04	0.62	0.77
% total solids	6.5	37.5	10.7	76.7
% fat	0.9	16.47	2.22	73.7
% protein	2.5	15.8	3.1	1.8
% lactose	2.7	4.1	4.2	1.2
% chlorides	0.17	0.5	0.17	ND
% ash	ND	1.26	0.54	ND

ND, not determined.

preparation of these products and/or poor bacteriological quality of the raw milk used for their manufacture. In addition to the indicator microorganisms, pathogens such as *Salmonella* sp., *Yersinia enterocolitica*, *Listeria monocytogenes*, and enterotoxigenic *Staphylococcus aureus* have been recovered mainly from samples of *lben* and *jben*. Although there are no epidemiological reports of outbreaks linking Moroccan traditional dairy products with diseases caused by these pathogens, their presence in these products indicates potential health hazards for consumers. Therefore, there is need to implement corrective procedures to eliminate or reduce this risk. This can be achieved by the use of heat-treated milk instead of raw milk and through the use of selected starter cultures for preparation of these products.

OBJECTIVES

The application of modern technology to Moroccan traditional dairy products aims to assure the following:

- Large-scale production of these products year-round by replacing raw milk with dry milk and butter oil. This will solve the problem of seasonality in Moroccan milk production.
- Production of dairy products with standardized chemical and microbiological composition so that their quality can be more easily controlled and standards for each product can be established.

TABLE 2 Average Microbiological Counts of Moroccan Traditional Dairy Products (cfu/g or ml)

Microorganism	*Lben*(4)	*Jben*(3)	*Raib*(3)	*Zabda*(2)
Streptococci	7.6×10^8	5.1×10^8	1.4×10^8	5.0×10^6
Lactobacilli	1.0×10^3	3.2×10^8	2.6×10^6	2.4×10^5
Leuconostocs	1.7×10^5	2.6×10^8	2.8×10^6	1.8×10^4
T. coliforms	5.0×10^4	4.3×10^5	1.7×10^5	6.5×10^4
F. coliforms	1.0×10^3	2.7×10^4	4.2×10^3	2.1×10^4
Enterococci	1.0×10^5	2.4×10^5	2.2×10^4	8.6×10^4
Fungi	8.5×10^2	2.3×10^6	2.3×10^4	ND
Total flora	2.9×10^9	8.2×10^8	3.5×10^8	4.6×10^7

ND: not determined.

• Elimination of massive contamination of these products and reduction of health hazards associated with these contaminations by using heat-treated milk and improving the sanitation and fermentation conditions.

• Adoption of simple and standardized processes for the preparation of these products that could be easily applied in the dairy industry.

PRELIMINARY STUDY

Preparation of traditional dairy products using improved technological processes requires, for each type of product, determination of the characteristics that constitute an excellent-quality product. For this purpose, samples of each product were evaluated by a gustatory panel. The best products were then analyzed to determine their physical characteristics, chemical composition, and microbiological profiles. The objective of the study was to assess the sensorial and compositional parameters (e.g., acidity, total solids) that the improved product should have to be acceptable to consumers.

Selection of Starters

Microbiological analysis of the different traditional fermented dairy products showed that an important proportion of their microflora was represented by lactic acid bacteria. Lactic streptococci were predominant in *lben*, *raib*, and *zabda*, while streptococci, lactobacilli, and leuconostocs were found in *jben* at almost the same average levels (10^8 cfu/g or ml) (colony forming units). From each product isolates from the predominant lactic flora were identified using biochemical tests. The principal species found in *lben*, *raib*, and *zabda* were *Streptococcus lactis*, and *S. diacetylactis*, while *S. lactis*, *Lactobacillus casei casei*, and *Leuconostoc lactis* were the main species recovered in *jben*.

Owing to the nature of traditional Moroccan dairy products (fresh fermented products), the major criterion considered for selection of lactic starters was their acid production ability at different incubation temperatures. Production by lactic strains of certain substances contributing to the overall aroma of these products also was taken into account. Thus, several lactic strains were retained to be used for preparating improved products.

Manufacture of Traditional Dairy Products from Heat-Treated Milk

To prepare each type of product, a simple and economically feasible technology, which industrial dairy plants could easily adopt, was used.

The improved processes proposed for use with *raib* (fermented milk) and *jben* (fresh cheese) are as follows:

- Manufacture of *raib*:
  Reconstitution of dry milk to 90 percent water and 10 percent solids.
  Pasteurization at 63°C for 30 minutes.
  Addition of calcium chloride and storage at 7°C for 10 hours.
  Addition of fresh pasteurized milk (60 percent of the total volume).
  Inoculation (*S. lactis, S. diacetylactis* @ 3.0 percent).
  Distribution into plastic containers and incubation at 30°C for 3 to 4 hours.
  Refrigeration at 4° to 6°C.
- Manufacture of *jben:*
  Reconstitution and pasteurization of powdered milk.
  Addition of calcium chloride and storage at 7°C for 10 hours.
  Addition of fresh pasteurized milk (60 percent of the total volume).
  Inoculation (*S. lactis, S. diacetylactis, L. casei casei* @ 3.0 percent).
  Storage of inoculated milk at 20° to 25°C until 0.25 percent lactic acid is formed.
  Addition of rennet (5 to 10 milliliters/100 liters).
  Fermentation at 20° to 25°C until 0.60 percent lactic acid is formed.
  Curd cutting and whey draining.
  Unmolding when titratable acidity reaches 0.9 percent lactic acid and total solids content reaches 28 to 30 percent.
  Cutting of cheese into suitable pieces (150 grams/piece).
  Surface dry salting, if desired (1 percent salt) and wrapping.

RESULTS

This study is still in progress. The final results regarding sensorial quality, chemical composition, and microbiological quality of traditional dairy products made with the improved technology are not yet available. Nonetheless, preliminary data obtained for *raib* and *jben* are very encouraging:

- Sensorial quality: Laboratory samples of improved *raib* and *jben* gave similar or even higher sensory scores than market samples. The characteristics considered in this evaluation are mainly acidity, texture, and aroma.
- Chemical composition; Because standard procedures were used for making *raib* and *jben*, the samples obtained had uniform composi-

tions. This information will be useful in establishing standards for these products.

• Microbiological quality: The use of heat-treated milk in the manufacture of *raib* and *jben* had a profound effect on the microbiological quality of the products. The improved products were free from pathogens such as *S. aureus, Salmonella, L. monocytogenis*, and *Y. enterocolitica*. They were either free of or contained very few coliforms (<10 cfu/g). Their microbiological quality was substantially improved compared with currently marketed traditional products.

CONCLUSIONS

Although data on all traditional dairy products are not yet available, information on the quality of laboratory-made *raib* and *jben* indicates that the use of modern technology in their manufacture has enhanced their bacteriological quality and reduced the risks of dairy-borne infections. This new technology has also begun to establish standards for these products.

In addition, the manufacture of traditional dairy products at an industrial scale will increase the production of these products and assure better distribution and marketing.

On the other hand, the use of dry milk, which is more economical than raw milk for preparing products such as *raib* and *jben*, has the advantage of being available any time of the year. This is very important in Morocco, where seasonal variabilities in milk production are a major problem for the dairy industry. Furthermore, the availability of dairy products that are rich in nutrients (e.g., proteins, fat) at a modest price and throughout the year will contribute to reduced malnutrition especially among children in rural areas.

REFERENCES

1. El Marrakchi, A.M., M. Berrada, M. Chahboun, and M Benbouhou. 1986. Etude chimique du smen marocain. Le Lait 66:117-133.

2. Hamama, A. 1989. Studies on the hygienic quality of certain Moroccan dairy products. Ph.D. thesis, University of Minnesota.

3. Hamama, A., and M. Bayi. 1991. Composition and microbiological profile of two Moroccan traditional dairy products: *raib* and *jben*. Journal of the Society of Dairy Technology.

4. Tantaoui Elaraki, A., M. Berrada, A. El Marrakchi, and A. Berramou. 1983. Etude du *lben* marocain. Le Lait 63:230-245.

11

Fermented Milk Products in Zimbabwe

Sara Feresu

Fermentation is the oldest means of preserving milk (1). Originally, unpasteurized milk was left to ferment naturally, and fermentation involved microorganisms present in the raw milk and surrounding air. With the development of modern technologies, specific lactic-acid-producing microorganisms are now introduced to carry out fermentations under controlled conditions. In this way fermented products of superior nutritional, physical, chemical, and sanitary qualities are produced.

In Zimbabwe one finds the modern fermented products such as yogurt and different types of cheese. The rural population, however, still ferment their milk traditionally. Fresh unpasteurized cow's milk is allowed to stand, at ambient temperature, in an earthenware pot loosely covered by a plate. This allows microorganisms inherent in the milk, from the pot, and from the surrounding air to ferment the milk. Fermentation takes 1 to 2 days depending on the ambient temperature (20 to 25°C). The fermented milk is not refrigerated and has an estimated shelf life of 3 days at ambient temperature.

In response to the urban population's desire for fermented milk, the Zimbabwe Dairy Marketing Board produces a fermented milk called Lacto on an industrial scale. Milk is standardized, pasteurized at 92°C for 20 minutes, cooled to 22°C, and inoculated with 1.2 percent of an imported mesophilic starter culture, similar to that used to produce "filmjolk," a Scandinavian fermented milk. The milk is immediately packaged into sachets, left to ferment at ambient temperature for 18 hours, and stored at 5°C ready for the market. The shelf life of refrigerated Lacto is 7 days.

Our studies have compared traditionally fermented milk with Lacto. We included traditionally fermented pasteurized milk, since substitution of unpasteurized with pasteurized milk might be an alternative for upgrading hygienic standards. The initial study was concerned with the effects of pasteurization and of the container used during fermentation on the total microbial cell counts, the counts of lactic acid bacteria, the amount of lactic acid produced, and the acceptability of the fermented milk by a panel (2).

We have characterized 10 predominant lactic acid bacterial isolates from traditionally fermented milk and four isolates from Lacto (3). We have also carried out studies to determine the fate of pathogenic and nonpathogenic *Escherichia coli* during fermentation of Lacto and traditionally fermented pasteurized and unpasteurized milk. The survival of *E. coli* was also tracked during storage of the fermented products at ambient (20°C) and refrigeration temperatures (5°C) for 4 days (4), since it is possible that pathogenic bacteria may gain access to these products before, during, and after fermentation. In the case of traditionally fermented milk, coliform contamination from cattle dung or from the milker's hands is possible. Contamination with coliforms during Lacto production can occur through bulk starter cultures and from inadequately sanitized equipment.

TRADITIONALLY FERMENTED MILK AND LACTO

In an earlier study (2) unpasteurized milk and pasteurized milk were fermented in clean nonsterile earthenware pots and sterile glass containers. At the same time, Lacto was fermented in plastic sachets and sterile glass containers. Bacterial counts and lactic acid levels were determined. The acceptability of the fermented milks was ranked by 11 panelists. Comparisons of all parameters were made after 24 and 48 hours of fermentation, when Lacto and traditionally fermented milk are likely to be consumed.

The numbers of lactic acid bacteria, lactic acid production, and acceptability were always higher for unpasteurized than pasteurized traditionally fermented milk irrespective of the container used.

Earthenware pots are better containers for traditional fermentation of milk. This is because earthenware pots have micropores in their walls, which, if not sterilized, may harbor lactic acid bacteria from the previous fermentation, which then act as inocula for the next fermentation. Our results suggest that earthenware pots are good containers to ferment milk in and may still have a place in milk fermentation in the home.

Unpasteurized milk fermented traditionally in either container was significantly more acceptable to the panel than Lacto, although the products were similar in all the other parameters assessed. It was therefore impossible to explain the differences in the acceptability of traditionally fermented milk and Lacto on the basis of this work. We suggested that the differences were probably due to the types of microorganisms involved in the fermentation of the two milk products rather than pasteurization or the container used for fermentation. Thus, we set out to isolate and characterize the lactic acid bacteria in traditionally fermented milk and Lacto.

ISOLATION AND IDENTIFICATION OF LACTIC ACID BACTERIA

From the previous study (3), 10 predominant morphologically different lactic acid bacteria colony types from plates inoculated with traditionally fermented milk and four morphologically different types of colonies from Lacto plates were selected and isolated into pure culture. The isolates were identified using numerical taxonomic techniques and reference strains. The isolates and reference strains were examined for 32 characteristics. Data were analyzed using the simple matching coefficient, and clustering was by unweighted pair group average linkage (5).

All the isolates from traditionally fermented milk belonged to the genus *Lactobacillus*. Seven of the isolates could be identified as belonging to *L. helveticus, L. plantarum, L. delbrueckii* subspecies *lactis* (two isolates), *L. casei* subsp. *casei* (two isolates) and *L. casei* subsp. *pseudoplantarum*. Three of the isolates could only be identified as either betabacteria or streptobacteria. The four isolates from Lacto were identified as *Lactococcus lactis*. They could not, however, be identified to subspecies level.

From this study we concluded that the differences in acceptability of traditionally fermented milk and Lacto are probably due to differences in the biochemical pathways and resulting types and levels of end products produced by the different bacteria responsible for fermentation of the two products. We suggested that more work should be done to determine the particular flavors and aroma present in traditionally fermented milk that are absent in Lacto and to determine whether any of our isolates are responsible for producing these desired properties.

E. COLI STRAINS IN LACTO AND TRADITIONALLY FERMENTED MILK

In another study (4) the growth and survival of pathogenic and nonpathogenic strains of *E. coli* were determined in traditionally fermented pasteurized and unpasteurized milk and Lacto. Unpasteurized and pasteurized milk and freshly inoculated Lacto, together with sterile control milk, were each inoculated with two strains of pathogenic and one strain of nonpathogenic *E. coli* to give approximately 10^3 cells/milliliter. All the milk treatments were left to ferment at ambient temperature (20°C) for 24 hours. One set of the fermented products was stored at ambient temperature, and the other set was refrigerated (5°C) for another 96 hours. Samples were taken at 24-hour intervals and tested for numbers of *E. coli*, pH, and percentage of lactic acid.

Lacto inhibited all three *E. coli* strains. Two strains (one pathogenic and one nonpathogenic) could not be recovered, and the third (pathogenic) survived only in very low numbers after 24 hours of storage of Lacto at both 20° and 5°C.

All three *E. coli* strains survived and multiplied to maximum cell numbers in the range 10^7 to 10^9/milliliter during traditional fermentation of unpasteurized milk. Cell numbers decreased to 10^3 to 10^6 and 10^2 to 10^5 during storage of the fermented product at 20° and 5°C, respectively. These results indicated that traditional methods of fermenting milk in Zimbabwe pose a potential health hazard because, if milk is contaminated during milking or fermentation, *E. coli*, and possibly other enteric pathogens, are able to multiply to infective doses and retain relatively high numbers during storage of the product at both refrigeration and ambient temperatures. The results also indicated that more than acid production alone is involved in the fate of *E. coli* during fermentation and storage of Lacto and traditionally fermented unpasteurized milk since more *E. coli* survived in unpasteurized fermented milk despite similar final lactic acid and pH levels of both milk products. We suggested that, since in our earlier studies we found that different lactic acid bacteria were responsible for fermentation of the two milk products, it is likely that these organisms produce different types and quantities of other inhibitory products (antibiotics, volatile acids, hydrogen peroxide) during fermentation.

Higher maximum numbers, 10^9 to 10^{10} of the three strains of *E. coli*, were attained during traditional fermentation of pasteurized milk. The numbers decreased to 10^5 to 10^8 and 10^4 to 10^7 during storage of the fermented product at 20° and 5°C, respectively. Under our experimental conditions there appeared to be more danger in traditionally fermenting pasteurized milk than unpasteurized milk; since less acid was produced, more *E. coli* multiplied and survived during fermentation and during storage of the pasteurized fermented milk. The practical relevance of this result should be interpreted with caution, since pasteurization also removes milk-borne organisms such as *E. coli* and *Salmonella* spp. and since it is unlikely that airborne recontamination of the milk by *E. coli* would result in initial numbers as high as 10^3 cells/milliliter. Thus, use of pasteurized milk in practice may not be as inappropriate as it might appear in theory.

Generally, fewer *E. coli* survived when the fermented milk products were stored at refrigeration than at ambient temperature. However, most people in rural areas of Zimbabwe do not have access to refrigerators.

CONCLUSIONS

We are currently determining the amounts of some B vitamins and of aroma and flavor compounds in traditionally fermented unpasteurized milk and Lacto. Preliminary results indicate that traditionally fermented milk contains more thiamine, riboflavin, pyridoxine, and folic acid than Lacto. Again, traditionally fermented unpasteurized milk is performing better than Lacto.

From the work we have done so far there are two options to follow in our future studies. We know that traditionally fermented milk has similar amounts of lactic acid and a pH level similar to that of Lacto and that it might also have higher amounts of some B vitamins; however, it is not hygienically acceptable. We know some of the lactic acid strains involved in the fermentation, but we also know that in a situation where raw milk is used and fermentation is carried out under conditions where asepsis is not observed, other microorganisms, in addition to lactic acid bacteria, contribute to the production of aroma and flavor compounds. Supposing we were to develop a starter culture based mainly on members of the genus *Lactobacillus*, it is debatable whether we would have the same organoleptic properties in a traditionally fermented pasteurized milk as found in traditionally fermented unpasteurized milk. If we developed and sold this starter culture for home use in fermentation of boiled milk, it is also unlikely that poor rural people would adopt such a fermentation since it has an added cost when compared with traditional fermentation.

Alternatively, we could incorporate some isolates from traditionally fermented milk into the Lacto starter culture and see whether the organoleptic properties of Lacto can be improved. Such a product would have to taste much better than traditionally fermented unpasteurized milk so as to entice rural populations to abandon traditional fermentation and adopt Lacto. Educational programs would have to be instituted for the public to appreciate the wisdom of spending money on buying Lacto, a hygienically safer product. At present, it is unlikely that Lacto will replace traditionally fermented milk in the foreseeable future.

REFERENCES

1. Robinson, R. K., and A. Y. Tamime. 1981. Microbiology of fermented milks. Pp. 245–278 in: Dairy Microbiology, Vol. 2. R. K. Robinson, (Ed.). London: Applied Science Publishers.

2. Feresu, S., and M. I. Muzondo. 1989. Factors affecting the development of two fermented milk products in Zimbabwe. MIRCEN Journal of Applied Microbiology and Biotechnology 5:349-355.

3. Feresu, S., and M. I. Muzondo. 1990. Identification of some lactic acid bacteria from two Zimbabwean fermented milk products. World Journal of Microbiology and Biotechnology 6:178-186.

4. Feresu, S., and H. Nyati. 1990. Fate of pathogenic and non-pathogenic *Escherichia coli* strains in two fermented milk products. Journal of Applied Bacteriology 69:814-821.

5. Sneath, P. H. A., and R. R. Sokal. 1973. Numerical Taxonomy: The Principles and Practice of Numerical Classifications. San Francisco: W. H. Freeman, pp. 228-234.

IV. PLANT DERIVATIVES

12

Cassava Processing in Africa

Olusola B. Oyewole

Cassava is an important food crop in the tropics and many countries in Africa. The crop contributes significantly to the diets of over 800 million people, with per capita consumption averaging 102 kilograms per year. In some areas of Africa it constitutes over 50 percent of the daily diets of the people.

Traditionally, cassava is processed before consumption. Processing is necessary for several reasons. First, it serves as a means of removing or reducing the potentially toxic cyanogenic glucosides present in fresh cassava. Second, it serves as a means of preservation. Third, processing yields products that have different characteristics, which creates variety in cassava diets.

The objective of this paper is to detail the strategy and program being followed in our laboratory to utilize the knowledge of biotechnology to improve the processing of cassava in Africa.

TRADITIONAL PROCESSING

Processing of cassava for food involves combinations of fermentation, drying, and cooking. Fermentation is an important method common in most processings. While there are many fermentation techniques for cassava, they can be broadly categorized into solid-state fermentation and submerged fermentation. Solid-state fermentation, typified by *gari* production, uses grated or sliced cassava pieces that are allowed to ferment while exposed to the natural atmosphere or pressed in a bag. Submerged fermentation involves the soaking of whole peeled, cut and peeled, or unpeeled cassava roots in water for various periods, as typified by the production of *fufu* and *lafun* in Nigeria. Traditionally, cassava is fermented for 4 to 6 days in order to

The support of the International Foundation for Science, Stockholm, Sweden, is gratefully acknowledged.

effect sufficient detoxification of the roots. Some processors, out of economic pressure, ferment cassava for less than 2 days. Some cases of food poisioning have been attributed to this practice. Application of biotechnology to traditional cassava processing has prospects for producing safe and well-detoxified products.

RESEARCH APPROACH

Our approach on cassava processing research is divided into three areas:

1. Investigating the science of the traditional submerged fermentation of cassava to *fufu* and *lafun* production;
2. Optimization of the processing through process controls; and
3. Improvement of traditional processing through application ot biotechnological techniques.

The microorganisms involved in the submerged fermentation process have been isolated and found to include *Bacillus subtilis*, *Klebsiella* sp., *Candida tropicalis*, *C. krusei*, and a wide spectrum of lactic acid bacteria, major among which are *Lactobacillus plantarum* and *Leuconostoc mesenteroides*. A microbial succession trend was found with the starch degrading *Bacillus subtilis*, giving way to the lactic acid bacteria and yeasts that dominate the latter part of the fermentation. The pH and titratable acidity of the fermenting cassava roots increased from 6.3 ± 0.2, 0.08 ± 0.03 percent to 4.0 ± 0.3, 0.36 ± 0.05 percent, respectively, after the 96-hour fermentation period. Organic acids produced during fermentation include lactic, acetic, propanoic, and butanoic acids, among others, and these are believed to contribute to the characteristic flavor of fermented cassava products. Fermentation causes release of some bound minerals, including calcium and magnesium. The most important contribution of fermentation is the release from the plant tissues of the enzyme linamarase, which is involved in the breakdown of the linamarin and lotaustralin (cyanogenic glucosides) of cassava, which releases hydrogen cyanide and thus detoxifies the product.

PROCESS OPTIMIZATION

Processing conditions for optimizing the fermentation process have been investigated in our laboratory. We found that the temperature range of 30° to 35°C, with a soaking period of 48 to 60 hours, is best for submerged processing. The size to which the roots are cut, peeling or nonpeeling of roots before processing, changing or nonchanging of fermentation water at intervals during processing, and the age of roots all affect the characteristics of the final product. In addition, the protein contents of products can be improved by cofermentation with legumes such as soybeans and cowpea.

BIOTECHNOLOGICAL INVESTIGATIONS

The overall goal of our biotechnological investigations is to develop an appropriate starter culture for cassava processing that will effectively produce linamarase enzymes for detoxifying cassava, break down starch to the simple sugars needed for acid production, improve the protein content of the products, reduce processing time, and yield products with stable desired qualities. The following summarizes our current findings:

- The microorganisms involved in fermentation have been characterized.
- Characterized isolates were used as single and multiple starter cultures for cassava fermentation. This has made it possible to understand the roles of each of the microorganisms implicated in the natural fermentation process. *Bacillus subtilis* and *Klebsiella* spp. contribute significantly to the rotting of cassava roots. In addition, *B. subtilis* produces amylase enzymes that are necessary for the breakdown of starch to sugars, which are needed for the growth of other fermenting microorganisms, including the lactics. Yeasts play a major role in odor development and, where high yeast biomass is encouraged, protein-enriched products are not. Lactic acid bacteria convert cassava sugars to lactic and other acids that contribute to the flavor in addition to having preservative effects.
- Appropriate starters have been developed that can produce amylase and linamarase enzymes necessary for starch breakdown and cyanogenic glucoside hydrolysis; two major biochemical processes needed in cassava processing. For this, the lactic acid bacteria were investigated since they were the predominant microbial group present at the beginning of fermentation and which persist and survive the acidic conditions that prevail in cassava fermentation. To date, we have found strains of *Lactobacillus plantarum* that are capable of producing amylase and linamarin. The linamarase produced has been purified, and it exhibits optimal activity at pH 5 to 8 and temperature of 30° to 40°C. Prospects for cassava processing using a selected single culture with properties for starch hydrolysis, cyanide detoxification, and acid production have thus evolved.
- To initiate genetic manipulation of cassava lactic acid bacteria,

the plasmid profiles of the lactobacilli isolated from cassava were studied. The presence of plasmids among cassava lactobacilli has been confirmed. Further research is needed to investigate the correlation between possession of plasmids and linamarase production in order to establish prospects for genetic manipulation.

FUTURE RESEARCH

Beyond the selection of appropriate starter cultures for cassava fermentation, it will be necessary to improve the starter culture. Genetic manipulation of the starter culture offers the best hope for improved cassava processing, with higher economic returns and improved stable qualities.

Cassava processing could also be enhanced by using biotechnological principles to modify structural and processing characteristics of cassava cultivars to meet specific product requirements.

The linamarase elaborated by cassava plant tissues and fermenting microorganisms has been found to be unstable under high acidic conditions characteristic of the latter part of natural fermentation. Techniques for increasing the stability of linamarase enzyme to acidic conditions could be investigated.

The usefulness of cassava fermenting microorganisms could be further investigated for the production of other economically viable products such as acidulants and antimicrobial agents.

A biotechnological approach could be investigated for the treatment of odorous fermented cassava water and cassava root peels.

13

Improving the Nutritional Quality of *Ogi* and *Gari*

T. G. Sokari

Ogi is a blancmange-like product processed by fermenting the slurry from wet-milled maize (or sorghum or millet). Used as both a weaning food for infants and as a breakfast food by adults, *ogi* is one of the most important food items in Nigeria. Yet it is nutritionally inferior to maize, which is deficient in certain essential amino acids, because of the maize-milling process that is an integral part of *ogi* production.

Cassava, another very important food crop, has the problem of possible nutritional complications because it contains the cyanogenic glucosides linamarin and lotanstralin. Although the cyanogens in cassava are hydrolyzed to hydrogen cyanide during processing by the endogenous enzyme linamarase (1,2), not all processes are equally effective. It has even been suggested that traditional processing techniques are unlikely to remove all the cyanide from cassava (3,4).

In view of this, studies were undertaken to increase the protein content of *ogi* relatively inexpensively and to develop a technique for processing cassava into *gari* that would eliminate cyanogens from the product or reduce them to innocuous levels.

PROCESSING OF *OGI*

The traditional technique for processing maize into *ogi* is summarized in Figure 1. Also shown in Figure 1 is an alternative to this procedure; a 20-minute boiling step is substituted for the normal 24- to 28-hour steeping of maize prior to wet milling (5). Cowpea can be combined with maize to increase the protein content of *ogi*.

Substituting 20 minutes of boiling for the traditional 24 to 28 hours of steeping prior to wet milling maize reduced processing time from 72-76 hours to about 24 hours. There was, however, no significant

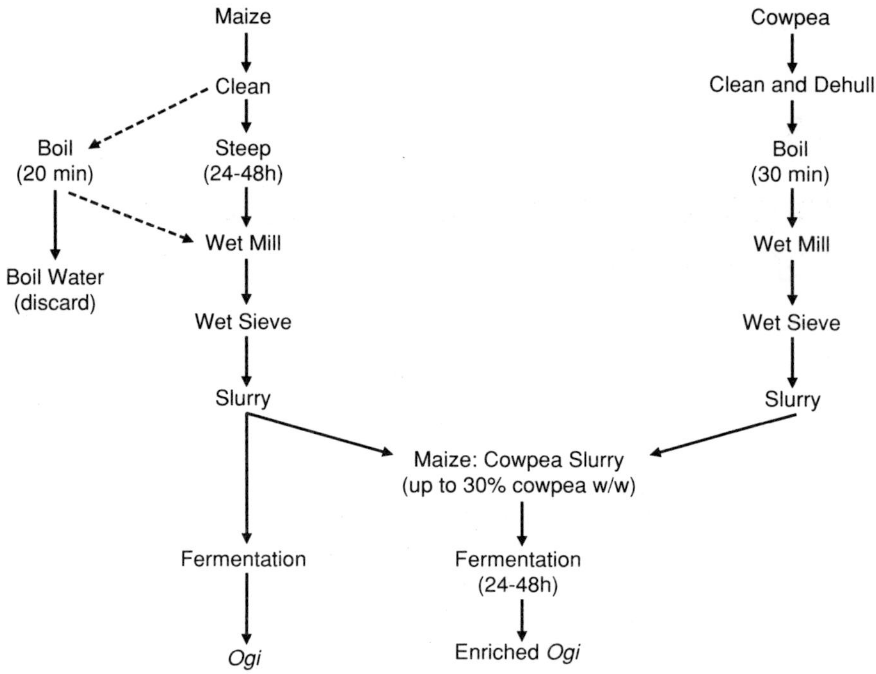

FIGURE 1 Processing of maize into *ogi* including cowpea protein enrichment; dotted lines show a bypass by which processing time can be reduced.

difference ($p > 0.05$) in the aroma, color, taste, and overall acceptability between the products obtained by the short-time processing and traditional processing (5). The same was also true for unenriched and protein-enriched *ogi* except for color (6).

CYANIDE REDUCTION DURING CASSAVA PROCESSING

Two foods processed from cassava (*gari* and *ijapu*) were studied. Adding water to grated cassava at the 75 percent (v/w) level and heating at 50°C for 6 hours resulted in linamarin reduction of >99 percent (Figure 2). The pH of the mash fell from 6.4 to 6.3 during the period (7). After dewatering, the mash was adjusted to a pH below 4 by equilibrating with a 3-day fermented cassava liquor (40 percent, v/w) at 50°C for 12 to 18 hours. The equilibrated mash was then dewatered and toasted (Figure 3). A panel of tasters who were familiar with *gari* but otherwise untrained could not differentiate between the product and traditionally processed *gari*. Both sets of products were equally acceptable to the panel.

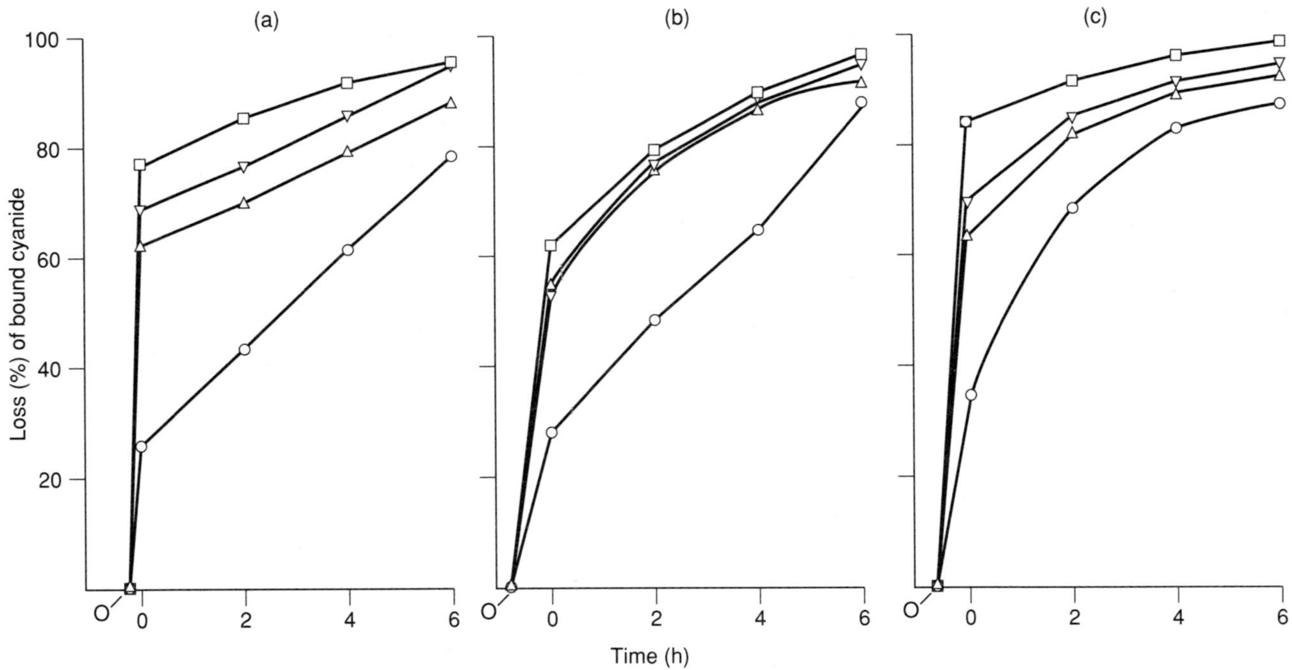

Figure 2 Effect of added water at 0%—i.e., control ○-○, 25% △-△, 50% ▽-▽, 75% □-□ levels on linamarin hydrolysis in grated cassava at (a) 30°C, (b) 40°C, (c) 50°C.

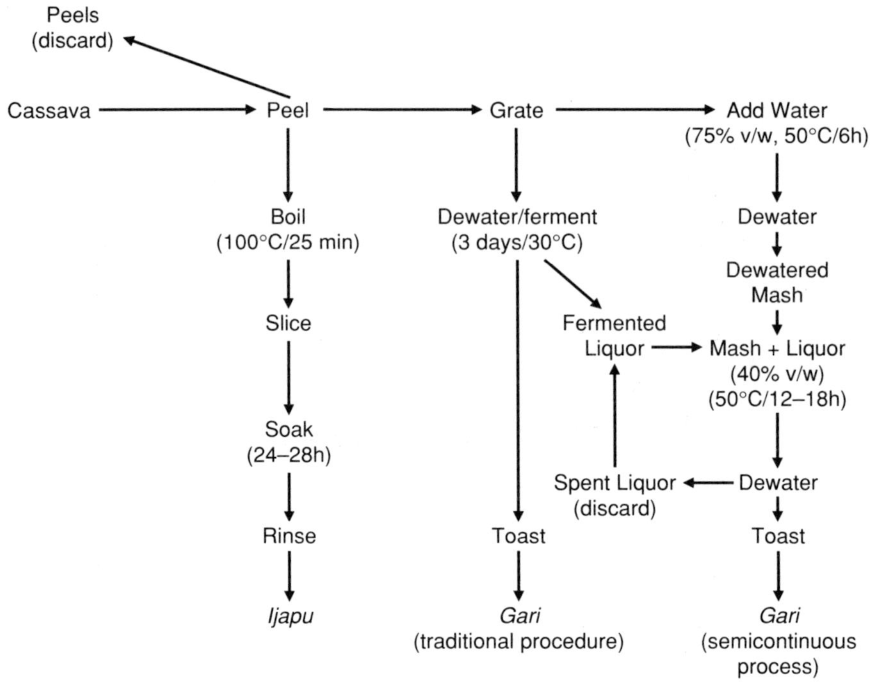

FIGURE 3 Processing of *ijapu* and *gari* from cassava.

The modified procedure for processing cassava into *gari* reduced the processing time from >96 hours to about 24 hours. Cyanogens were not detectable in the product by the method of Cooke (8,9).

The study of the traditional production of *ijapu* (Figure 3) was intended to aid in understanding the loss of cyanogens during cassava processing. About 54 percent of the cyanogens in raw cassava were lost after boiling peeled cassava, but a substantial proportion remained in the water (Table 1). After slicing the boiled cassava and steeping the slices, a substantial proportion of the cyanogens was again lost.

TABLE 1 Cyanide Content of Cassava During Processing Into *Ijapu*

		Cyanide Content (ppm)				
Material analyzed	pH	Total	Free	HCN	Bound	Cyano
Unprocessed peeled cassava	6.3	76.1±15.3	5.5±2.2	2.9±0.4	70.6	2.6
Boiled cassava	6.0	35.1±8.7 (53.9)	2.4±1.2	2.0±0.7	32.7 (53.7)	0.4
Boil water	6.2	12.8±2.1	1.1±0.2	0.5±0.2	11.6	0.6
"Ijapu" (after 24 hours steeping)	ND	11.8±1.4 (84.5)	4.2±1.4	3.1±1.4	7.6 (89.2)	
Steep water	4.0	16.2±3.9	5.7±3.4	5.6±3.4	7.8	0.1

Note: Cyano, Cyanohydrin; numbers in parenthesis, percent loss; ND, not determined

Much of the loss could, however, be accounted for in the steep water, and the proportion lost depended on the duration of steeping and the cassava:water ratio (Tables 1 and 2).

ROLE OF FERMENTATION

The reduction of >99 percent in the linamarin content of grated cassava within 6 hours of adding water, with little or no change in the pH of the mash, would imply that fermentation had nothing to do with the detoxication. Linamarin breakdown is essentially a hydrolytic process catalyzed by the endogenous enzyme linamarase (1,2). The results of the present study indicate that the addition of water aids in the hydrolytic process. Apparently not all of the water normally in raw cassava tuber is available for hydrolysis.

During the boiling of cassava for processing into *ijapu*, linamarase would be inactivated. Yet a substantial proportion of the linamarin in cassava was still lost, appearing to a large extent in the water used for boiling and for steeping (Tables 1 and 2). This would suggest that leaching could be an important factor in cyanide loss during cassava processing. This would be true not only during the boiling and steeping of cassava for *ijapu* production but also during the dewatering of grated cassava for *gari* production.

CONCLUSION

Cassava detoxication during processing is essentially an enzymic hydrolysis of cyanogens in cassava (1,2,8). Fermentation has little role

TABLE 2 Effect of Sliced Boiled Cassava: Steep Water Ratio on Cassava Detoxification

Ratio (w/v)	Material analyzed	pH	Total	Free	HCN	Bound	Cyano
1:1	Unboiled cassava	6.4	90.2	4.5	3.0	85.7	1.5
	Boiled cassava[a]	ND	51.7 (42.7)	1.5	0.5	50.3 (41.3)	1.0
	Boil water	6.1	20.2	0.5	0.3	19.7	0.2
	Sliced, boiled cassava[b]	ND	5.6 (93.8)	1.1	0.6	4.5 (94.7)	0.6
	Steep water	4.3	25.5	5.7	4.5	19.7	1.3
1:2	Sliced boiled cassava[b]	ND	5.7 (93.7)	1.7	0.6	4.0 (95.3)	1.2
	Steep water	4.3	21.3	1.9	1.4	19.4	0.5
1:3	Sliced, boiled cassava[b]	ND	4.2 (95.3)	1.6	0.5	2.6 (97.0)	1.0
	Steep water	4.2	11.5	2.1	1.1	9.4	1.0

[a] Prior to slicing and soaking in water.

[b] After 24 hours soaking; other notes as in Table 1.

in this process and may even be antagonistic to it (10). Leaching is another important process for cyanogen reduction during cassava processing. Although fermentation does not aid in cassava detoxication during processing, it is important in flavor development (11,12) and preservation of the product.

REFERENCES

1. Conn, E. E. 1969. Cyanogenic glucosides. Journal of Agricultural and Food Chemistry 17:519–526.

2. Nartey, F. 1978. Cassava: Cyanogenesis, Ultrastructure and Seed Germination. Copenhagen, Denmark: Munksgaard.

3. Cooke, R. D., and E. N. Maduagwu. 1978. The effects of simple processing on the cyanide content of cassava chips. Journal of Food Technology 13:299–306.

4. Oke, O. L. 1983. Processing and detoxication of cassava. Pp. 329–336 in: Proceedings: 6th Symposium of the International Society for Tropical Root Crops, Lima.

5. Sokari, T. G., P. S. Karibo, and L. F. F. Manuel, 1991. Substitution of boiling for steeping in *ogi* production. Discovery and Innovation (In press).

6. Manuel, L. F. F. 1990. Microbiological, Nutritional and Sensory Evaluation of the Effects of Cowpea Fortification of *Ogi*. M.Phil. thesis, Rivers State University of Science and Technology, Port Harcourt, Nigeria.

7. Sokari, T. G., P. S. Karibo, and C. K. Wachukwu. 1991. Reevaluation of the role of fermentation in cassava detoxification during processing into foods. Proceedings: Workshop on Traditional African Foods, Dar-es-Salaam, Tanzania (In press).

8. Cooke, R. D. 1978. An enzymatic assay for the total cyanide content of cassava. Journal of the Science of Food and Agriculture 29:345–352.

9. Cooke, R. D. 1979. Enzymatic assay for determining the cyanide content of cassava and cassava products. Cassava Information Center, Centro Internacional de Agricultura Tropical, Cali, Colombia, 05EC-6, 14 pp.

10. Maduagwu, E. N. 1983. Differential effects on the cyanogenic glucoside content of fermenting cassava activities. Toxicology Letters 15:355–359.

11. Vasconcelos, A. T., D. R. Twiddy, A. Westby, and P. J. A. Reilly. 1990. Detoxification of cassava during *gari* preparation. International Journal of Food Science and Technology 25:198–203.

12. Dougan, J., J. M. Robinson, S. Sumar, G. E. Howard, and D. G. Coursey. 1983. Some flavoring constituents of cassava and of processed cassava products. Journal of the Science of Food and Agriculture 34:874-884.

14

Solid-State Fermentation of Manioc to Increase Protein Content

Nguyen Ngoc Thao and Nguyen Hoai Huong

Manioc (cassava) is grown extensively in Vietnam and other tropical countries for its high yields in infertile soil. Although manioc is high in carbohydrates, its use is limited by its low protein content (1 to 4 percent). Manioc has been used at levels of 10 to 15 percent in poultry feed and 35 to 50 percent in pig feed. Powdered dried fish debris (gills, scale, tail, etc., from the fish processing industry), oil cake (from coconut or peanut oil production), or soybean flour have been used to raise protein levels in such feeds, but these products raise the price of feed significantly.

To upgrade the protein content in manioc, yeast cells or fungi can be inoculated in a manioc-containing medium along with nutrients containing nitrogen, phosphorus, and potassium. The use of mycelial fungi has the following advantages:

- The protein content in fermented product can increase to 30 percent.
- Fungal protein can be substituted completely for animal protein.
- The product has a low nucleic acid content.
- The product contains a favorable spectrum of amino acids.

Solid-state fermentation and liquid-state fermentation are two methods used for cultivation of fungus. Liquid-state fermentation processes are well developed in industrialized countries but are not suitable for rural farms in developing countries. Solid-state fermentation is a simple process that does not require modern equipment, power supply, or sterile conditions. In addition, the capital investment is low, permitting countryside operation and the use of available manual labor.

Many studies of solid-state fermentation of manioc have been conducted. The cultivation of *Aspergillus niger* in a manioc medium at 35° to 40°C for 30 hours has resulted in protein content increases of 5 to 18 percent; carbohydrate content decreased from 65 to 28 percent (1). The protein from this fermentation can be competitive with soybean protein.

In addition to *A. niger*, other fungi such as *A. awamori*, *A. hennebegii*, *A. fugamitus*, *Rhizopus chinensis*, and *Sephalo sporium lichlorniae* can be grown in acid medium at high temperatures. The protein content of the fermented product can reach 48 percent.

In Vietnam, *A. niger* and *A. hennebegii* were cultivated on a maltolized-manioc medium or a mixture of manioc and rice flour. This research comes from the demand of the husbandry industry and is designed to develop a fermentation process for on-farm use.

MATERIALS

Dried manioc pieces were ground to the size of 5 to 10 mm. Spores of *A. niger* were cultivated by surface fermentation on a medium containing rice hulls, rice bran, or manioc flour as carbohydrate, and urea (2 percent), ammonium sulfate (8 percent), and potassium phosphate (4 percent) at pH 4.5. Spores were collected after 7 days of cultivation.

METHODS

After a defined period of fermentation, the product was dried at 65° to 70°C, ground, and analyzed. The moisture content was determined by drying at 105°C to constant weight. The protein content was determined by precipitating with a solution of $CaSO_4$ (6 percent) and NaOH (1.25 percent); the precipitate was analyzed by the Kjeldahl method. The starch content was determined by hydrolyzing the preparation with HCl and using the Bertrand method. The reduced glucose content was determined by the Bertrand method.

Table 1 shows that *A. niger* could not grow in medium containing urea as the only nitrogen at a concentration of 4.5 percent because of the resultant alkalinity. With $(NH_4)_2SO_4$ as the N source, the pH was maintained at 4 to 5 during the fermentation. The maximum protein content was attained in medium containing urea (4 percent) and ammonium sulfate (5.8 percent). The content of protein can reach 17 percent in comparison with the one cultivated in only urea-containing medium. However, 1.55 percent N protein was achieved in the culture medium containing 3.1 percent N with the transformation efficiency of

TABLE 1 Effect of Nitrogen Sources on Protein Formation

N°	N Source	N Percent in Culture Medium	Protein Percent in Fermented Product	N Protein in Fermented Preparation
1 Urea	2.0	0.93	5.2	0.83
	3.3	1.54	8.0	1.28
	4.0	1.86	8.2	1.3
	4.5	2.10	no growth	—
	5.0	2.33	no growth	—
2 Ammonium sulfate (AS) 7.4		1.54	4.4	0.70
Urea 2% + AS 10%		3.05	7.2	1.52
Urea 3.3% + AS 4%		3.10	9.36	1.49
Urea 4.0% + AS 5.8%		3.1	9.7	1.55
Urea 4.5% + AS 7.4%		3.1	no growth	

49 percent. The transformation efficiency was 70 percent in medium containing only urea (4 percent).

The P and K elements (Table 2) were added to the medium containing urea 3.3 percent and ammonium sulfate 4.4 percent (1–5) or urea 3.3 percent (6–9), respectively. The results suggested that the P and K sources had no clear effect on protein formation.

In Table 3, the effect of humidity on the protein content is shown. Table 4 illustrates the effect of sterilizing conditions on the yield of protein. Table 5 shows the effect of the amount of inoculum culture on protein synthesis.

RESULTS

Manioc flour cannot be used as a carbohydrate source in the culture medium because it agglomerates and excludes air necessary for the growth of the fungal mycelium.

Manioc pieces of 0.5 to 1.0 centimeters are best for this solid fermentation method. The protein content of fermented preparation decreased 50 percent when using manioc pieces that were 1.0 to 2.0 centimeters in size.

The analysis of a fermented preparation after 2 days of fermentation, drying at 65° to 70°C, and grinding is shown in Table 6.

TABLE 2 Effects of Nutrients on Biosynthesis of Protein (The P and K elements were from chemical fertilizers)

Chemical Fertilizer P Percent + K Percent	Protein Percent	Chemical Fertilizer P Percent + K Percent	Protein Percent
1 3.3 + 1.0	10.21	6 3.3 + 1.0	10.0
2 2.3 + 0.5	11.34	7 2.3 + 0.5	10.0
3 2.3 + 1.5	10.41	8 1.3 + 0.5	11.25
4 4.3 + 0.5	11.46	9 0.3 + 0.5	9.62
5 4.3 + 0.5	9.3		

TABLE 3 Effect of Initial Humidity on Protein Content

N°	Humidity of Culture Medium[a]	Protein Percent	Notes
1	45	7.9	The change of humidity from 60 to 70 percent occurred depending on the atmospheric temperature and humidity.
2	50	9.36	
3	55	9.64	
4	60	11.08	
5	65	11.23	
6	70	11.37	
7	75	Poor growth	

[a] The medium for this experiment contained urea (4 percent), P (1.3 percent), and K (0.5 percent).

TABLE 4 Effect of Sterilization Conditions on Protein Production

N°	Temperature, °C	Time, Minutes	Percent Protein	Notes
1	100	45	8.6	The culture medium can be sterilized at 100°C in 45 minutes.
2	100	90	7.35	
3	120	30	7.6	
4	120	45	7.50	
5	120	60	6.30	

TABLE 5 Effect of the Amount of Inoculum Culture on Protein Synthesis

N°	Percent of Inoculum Culture	Percent Protein	Notes
1	0.5	6.0	The maximum percent of protein was achieved in 2 percent of inoculum culture. There was formation of black spore in fermented preparation when using more than 2 percent of inoculum culture.
2	1.0	8.6	
3	1.5	10.0	
4	2.0	12.5	
5	3.0	12.0	

TABLE 6 Product Analysis

N°	Index	Manioc Pieces Culture-Medium, Percent	Fermented Preparation, Percent
1	Protein	1.5–2.0	10.0–13.3
	Starch	33	11
	Reduce sugar	4.4	8.11
	Total sugar	8.4	13.00

CONCLUSION

This solid-state fermentation method can be used to upgrade by six to seven times the protein content in manioc pieces. The resulting fermented product contains 10 to 13 percent protein, which is suitable for use as a feed additive.

REFERENCE

1. Raimbault, M. J. 1985. Fermentation Technology 63(4):395–399.

15

Leaf and Seed Fermentations of Western Sudan

David B. Harper and M. A. Collins

Kawal, *sigda*, and *furundu* are fermented foodstuffs indigenous to the Kordofan and Darfur provinces of Western Sudan. All are produced by solid state fermentation of readily available plant materials of little or no economic value which, though unpalatable in its natural state (and indeed toxic in the case of *kawal*), contain protein rich in sulphur amino acids. In each case, fermentation yields a product that is not only organoleptically acceptable but also sufficiently highly regarded nutritionally by the local people to be employed as a meat substitute. As the sun-dried food can be stored indefinitely without deterioration, these fermentation processes represent a food preservation technique particularly well suited to the climate and conditions of this part of Africa. Biochemical and microbiological aspects of these fermentations and their nutritional implications have been investigated by Dirar (1), Dirar et al.(2), and Elfaki et al.(3).

PREPARATION

Kawal is prepared from the fresh leaves of a wild and reputedly toxic legume, *Cassia obtusifolia,* which are pounded to a paste and packed into an earthenware zeer buried to the neck in the ground in a shaded location. A layer of green sorghum leaves is placed on the surface of the paste and the zeer fitted with a lid that is sealed with mud. At intervals of 3 days the vessel is opened, the sorghum leaves removed, and the paste remixed thoroughly by hand. The repacked paste is covered with fresh sorghum leaves and the zeer resealed. After 11 to 15 days, the strongly smelling black mass is removed, molded into small balls, and dried in the sun for 5 days. The dried *kawal* is usually consumed in a stew with onions, okra, or other local vegetables.

The seedcake remaining after oil extraction from *Sesame indicum* seed is the raw material for the *sigda* fermentation. The bitter, indigestible seedcake made from nondecorticated seed is often used only as animal feed. In the traditional *sigda* process the seedcake is ground to a paste with warm water. *Kambo,* a local form of potash from the dried leachate of the ash of the central stems of the sorghum seed head, is frequently, but not invariably, added (3 to 20 g/kg). The mixture is packed in an earthenware vessel sealed with a cotton cloth and a close-fitting lid to minimize access of air. The fermentation lasts 3 to 7 days at ≅ 30 C° with occasional remixing, addition of water if necessary, and resealing of the container, after which the product is molded into small balls and sun-dried. Like *kawal*, *sigda* is usually consumed in a vegetable stew. A similar fermented food, *furundu*, is prepared from the crushed seeds of *karkade (Hibiscus sabdariffa)* by a process almost identical to that employed for *sigda*.

MICROBIOLOGY

The microflora of *C. obtusifolia* leaves (the substrate for the *kawal* fermentation) was dominated by four bacterial species, *Bacillus subtilis, Lactobacillus plantarum, Propionibacterium* sp., and *Staphylococcus sciuri*, and two yeasts, *Candida krusei* and *Saccharomyces* sp. Although the relative proportions of these organisms changed, all persisted in detectable numbers throughout fermentation. The principal species present during fermentation were *B. subtilis* and *Propionibacterium* sp., the other organisms comprising a comparatively small proportion of the population. No marked interspecial successional pattern occurred during fermentation.

The microflora of unfermented sesame seedcake was dominated by two bacterial species, *Pediococcus* sp. and *Streptococcus* sp., and two yeasts, *Saccharomyces* sp. and *Candida* sp. *Pediococcus* sp. was eliminated after the second day of fermentation, and the occurrence of the two yeasts was confined to the first half of the fermentation period. However, the homofermentative lactic acid bacterium *Streptococcus* sp. dominated the microflora throughout most of the fermentation. Additionally, the yeasts *Debaryomyces* sp. and *Torulopsis* sp. appeared in low numbers late in fermentation.

No detailed examination of the microflora during fermentation of *furundu* has been attempted, but the principal organism present in the final product was identified as a *Bacillus* sp.

PROTEIN CONTENT AND QUALITY

The crude protein content decreased only slightly, if at all, during fermentation of each substrate, indicating little loss of nitrogen during the process (Table 1). It is clear that the high sulphur amino acid content of all the fermentation substrates is largely retained in the fermented products, which compare favorably with the FAO reference protein in this respect (Table 2). The branched chain amino acids valine, leucine, and isoleucine also tend to be at a higher level in the protein of *sigda* and *furundu* than in the protein of their respective substrates. The other noteworthy feature is the markedly enhanced concentration of alanine in *sigda* and, to a lesser extent, in *furundu*, compared with the unfermented substrate. This increase is probably attributable to the transamination of pyruvate formed by oxidation of the lactic acid produced in the fermentation. Significantly, alanine concentration did not rise during the *kawal* fermentation where lactic acid production is negligible.

The overall protein quality of each of the fermented foods is determined by the content of lysine, which is limiting in the raw material for both the *sigda* and *furundu* fermentations and does not increase appreciably during fermentation. Nevertheless, the proteins of *kawal* and *furundu*, with chemical scores of 73 and 80, are of surprisingly good quality, whereas that of *sigda*, with a chemical score of 33, is no poorer nutritionally than the protein of the local staple cereal, sorghum.

MINERAL, CRUDE FIBRE, AND OIL CONTENT

Ash content of all fermented foods showed a substantial increase on that of the unfermented substrate, which, in part, reflects the mineral contribution made by clay scraped from the interior of the fermentation vessel during preparation but also, in the case of *sigda* and particularly *furundu*, the liberal addition of *kambo* (Table 1). The latter consists largely of potassium bicarbonate with smaller quantities of potassium chloride, silicate, and sulphate. *C. obtusifolia* leaves display an unusually high calcium content, which is believed to be critical in determining the course of fermentation (see below). Oil and crude fibre contents of the fermented foods was not significantly different from that of the unfermented substrates, suggesting that participation of these fractions in the fermentation process is unlikely.

CHANGES DURING FERMENTATION

The dominant role of lactic acid and the marked decrease noted in pH during the *sigda* fermentation contrast strongly with the high concentrations of volatile fatty acids (VFA) and minimal pH change observed in the *kawal* fermentation (Table 3 and Figure 1). Thus, by

TABLE 1 Composition of Field Collected *Kawal*, *Sigda*, and *Furundu* Compared With That of Their Fermentation Substrates on a Dry Weight Basis

	Ash %	Crude protein %	Oil %	Crude fibre %	K %	Na %	Ca %	Mg %	P %	S %	Fe $mg\ kg^{-1}$	Zn $mg\ kg^{-1}$	Mn $mg\ kg^{-1}$	Cu $mg\ kg^{-1}$
Leaves of *C. obtusifolia*	12.6	24.3	2.5	13.5	ND[a]	ND[a]	3.85	0.30	0.26	ND	534	32	75	ND
Kawal	19.6	26.2	3.8	12.1	2.5	ND[a]	4.13	0.42	0.28	0.52	82	84	112	11
Sesame seedcake[b]	14.0	45.6	14.4	7.4	1.04	<0.01	1.87	0.66	1.12	0.74	708	117	68	38
Sigda	18.2	43.8	16.9	8.2	1.83	0.67	2.25	0.66	1.11	0.75	509	127	83	34
Karkade seed	6.2	32.6	21.1	25.1	1.27	<0.01	0.31	0.43	0.62	0.36	313	90	118	18
Furundu	22.8	26.5	23.3	26.5	5.65	0.08	0.58	0.69	1.08	0.63	347	116	122	21

[a] ND, Not determined
[b] Assara extracted

TABLE 2 Amino Acid Composition of *Kawal*, *Sigda*, and *Furundu* and Their Fermentation Substrates Compared With the FAO Reference Protein

	Amino acid concentration (g 16 g^{-1} N)																		
	Asp	Thr	Ser	Glu	Pro	Gly	Ala	Val	Cys	Met	ILeu	Leu	Tyr	Phe	His	Lys	Arg	Orn	α - NH_2 But	γ - NH_2 But
Leaves of *C. obtusifolia*	12.1	6.2	4.6	13.6	7.7	6.7	7.5	7.5	1.4	2.1	6.0	10.4	5.3	6.8	3.3	7.7	7.2	<0.1	<0.1	2.6
Kawal	7.7	3.3	2.8	8.2	4.2	5.0	6.8	6.4	1.2	1.5	5.1	8.3	3.5	5.4	2.0	4.0	4.0	0.1	0.7	4.1
Sesame seedcake	7.8	3.1	3.7	20.9	3.9	5.4	4.7	6.3	1.6	2.4	3.2	6.6	3.2	4.2	2.4	2.0	12.8	<0.1	<0.1	<0.1
Sigda	7.7	2.6	3.1	20.1	4.2	6.0	9.9	6.0	2.1	2.5	4.6	8.0	2.8	4.5	2.0	1.9	10.4	<0.1	1.2	4.1
Karkade seed	11.0	3.2	4.8	24.3	4.1	5.4	4.4	4.5	2.2	3.1	3.4	6.8	3.2	4.8	2.4	4.2	13.0	<0.1	<0.1	<0.1
Furundu	10.4	3.5	3.5	20.2	4.5	5.9	6.2	5.3	1.9	2.6	3.6	7.1	2.6	4.3	2.0	4.4	7.7	0.9	<0.1	0.2
FAO reference protein		4.0						5.0	3.5		4.0	7.0	6.0			5.5				

TABLE 3 Lactic Acid and Volatile Fatty Acid Content of Field Collected *Kawal*, *Sigda*, and *Furundu* (Mean and Range in g 100 g^{-1} Dry Matter)

Acid	*Kawal*	*Sigda*	*Furundu*
Lactic	0.21 (0.03–0.51)	3.07 (2.85–3.35)	0.50 (0.03–1.67)
Acetic	5.08 (2.12–6.75)	1.10 (1.00–1.19)	1.59 (1.22–2.05)
Propionic	0.90 (0.51–1.59)	0.04 (0.03–0.05)	0.09 (0.02–0.25)
Isobutyric	0.24 (0.04–0.38)	<0.01	<0.01
n-Butyric	2.94 (1.18–4.73)	0.08 (0.02–0.15)	0.24 (0.05–0.73)
Isovaleric	0.22 (0.06–0.60)	0.02 (0.01–0.03)	0.17 (0.02–0.35)
n-Valeric	0.18 (0.01–0.61)	<0.01	0.01 (0.01–0.02)
Total VFA	9.56 (4.70–12.1)	1.24 (1.12–1.38)	2.17 (1.65–2.63)

the eleventh day of the latter fermentation, VFA—mainly n-butyric (8 percent), acetic (5 percent) and n-propionic (2 percent)—comprised 15 percent of the fermentation mixture. However, the pH had not changed by more than 0.5 unit from the initial value. On the other hand, by the fifth day of the *sigda* fermentation, when a total acid concentration of 6 percent had been attained, the pH of the fermentation mixture had fallen to 4.0 from an initial value of about 6.0. This difference in the course of fermentation is almost certainly attributable to the stronger buffering capacity of the substrate of the *kawal* fermentation, *C. obtusifolia* leaves, which possess approximately double the calcium content of sesame seedcake. Conditions in *kawal* do not, therefore, favor the selection of acidoduric lactic acid bacteria.

In addition to these bacteria, the two yeast species present in unfermented sesame seedcake proliferated during the initial period of fermentation. Concomitantly, starch levels were observed to fall rapidly from 2 percent in the unfermented substrate to zero after the first two days of fermentation. As the only amylolytic organisms present, the yeasts presumably were responsible for degradation of starch, rendering it available to the lactic acid bacteria. The poorly fermentative yeasts *Torulopsis* sp. and *Debaryomyces* sp. isolated in the final stages of the fermentation can utilize lactic acid aerobically and may cause the decline in concentration of the compound during this period. The addition of *kambo* did not appear to have any significant effect on the course of the *sigda* fermentation, and it was concluded that this supplementation was probably practiced mainly on organoleptic grounds.

The VFA, primarily n-butyric and acetic acids, which are the principal products of the *kawal* fermentation, are characteristic of clostridial fermentation of plant material as is the accumulation of

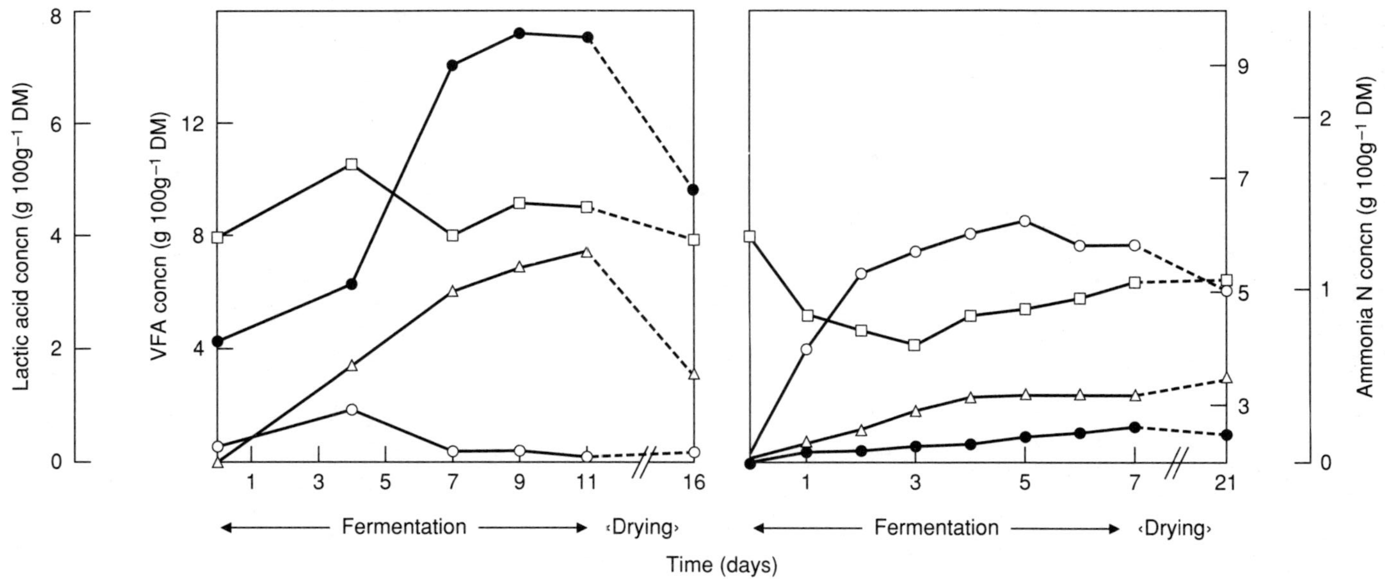

FIGURE 1 Changes in pH, VFA, lactic acid, ammonia-N concentrations during fermentation of (a) *kawal* and (b) *sigda*.

—□—, pH; —●—, VFA concentration; —○—, lactic acid concentration; —△—, ammonia-N concentration

ammonia nitrogen. However, all attempts to isolate clostridial species from the fermentation mixture were unsuccessful, indicating that the microbial origin of VFA must be sought elsewhere. The formation of acetic acid can probably be ascribed to the heterofermentative *B. subtilis,* the co-dominant microorganism, whereas propionic acid is probably an end product of anaerobic fermentation by *Propionibacterium* sp. for which lactate is a preferred substrate. Utilization of lactate in this way could explain the low level of lactate in *kawal*, despite the substantial population of *Lactobacillus plantarum*. The microbial pathway leading to formation of n-butyric acid is difficult to define, although its production may be characteristic of fermentation by this type of mixed culture as a whole rather than that by any single microorganism. n-Propanol (2.3 percent), n-butanol (0.1 percent), and ethanol (0.1 percent) were also detected in *kawal* toward the end of fermentation, though all were lost from the product during the drying phase. Formation of such alcohols is probably due to anaerobic fermentation of carbohydrate by the yeast species present.

The identification of *Bacillus* sp. as the principal microorganism in *furundu* when considered in the context of a final pH of 6.2 and the presence of both VFA and lactic acid in the fermentation mixture suggest that the *furundu* fermentation may be intermediate in character between those of *sigda* and *kawal*. Further investigation of the *furundu* fermentation would be most instructive in this respect.

CONCLUSIONS

The *sigda* and *furundu* fermentations appear quite unlike the traditional oilseed fermentations practiced in Nigeria and elsewhere in West Africa where foods such as *ogili* and *ogiri* are fermented from castor oil seed *(Ricinus communis)* and melon seed *(Citrullus vulgaris)*. There are even variations of the fermentation which use sesame seed and karkade seed known as *ogiri-sara* and red sorrel, respectively. During these West African fermentations, the pH increases to over 9.0 and ammonia production is frequently observed in the later stages. The fermentations are dominated by *Bacillus* sp., frequently *Bacillus subtilis,* an organism associated with spoilt *sigda* in Sudan. The principal reason for the difference would appear to be in the preparation of the seeds prior to fermentation, which in West Africa involves boiling for several hours in water until soft. Such pretreatment may alter the course of fermentation by two mechanisms—first, by rendering protein and polysaccharide more available for degradative attack by microorganisms, and second, by effectively eliminating much of the heat-sensitive indigenous microflora. The removal of amylolytic yeasts may well favor the selection of amylase-producing bacteria such as *Bacillus* sp. rather than lactic acid bacteria incapable of utilizing starch.

The three fermentations studied appear to afford a route by which unpalatable plant material or oilseed cake of little economic value can be converted into acceptable meaty-tasting food that is particularly rich in sulphur amino acids, which tend to be deficient in diets where access to meat or fish is limited. Phytic acid present in seeds can frequently hinder absorption of minerals in the gastrointestinal tract. As fermentation of plant products has been shown to reduce phytic acid levels substantially, it is likely that the bioavailability of minerals in both sesame and karkade seed is increased in *sigda* and *furundu*.

REFERENCES

1. Dirar, H. A. 1984. *Kawal* meat substitute from fermented *Cassia obtusifolia* leaves. Economic Botany 38:342–349.

2. Dirar, H. A., D. B. Harper, and M.A. Collins. 1985. Biochemical and microbiological studies on *kawal*, a meat substitute derived by fermentation of *Cassia obtusifolia* leaves. Journal of the Science of Food and Agriculture 36:881–892.

3. Elfaki, A. E., H. A. Dirar, M. A. Collins, and D. B. Harper. 1991. Biochemical and microbiological investigations of *sigda*—a Sudanese fermented food derived from sesame oilseed cake. Journal of the Science of Food and Agriculture 57.

16

Continuous Production of Soy Sauce in a Bioreactor

Takashi Hamada, Yaichi Fukushima, and Hiroshi Motai

Soy sauce is a traditional all-purpose seasoning with a salty taste and sharp flavor. In the conventional method of brewing soy sauce (Figure 1), cooked soybeans and roasted wheat are mixed with spores of *Aspergillus* species and fermented in solid culture for 2 days to produce *koji*. The *koji* is then mixed with brine to make *moromi*, the mash that ferments to produce soy sauce. Over time the soybeans and wheat are hydrolyzed by enzymes such as proteinases, peptidases, and amylases. During the first stage of *moromi* fermentation, *Pediococcus halophilus* grows and produces lactic acid, which lowers the pH. Accompanying the decrease in pH, vigorous alcohol fermentation by *Zygosaccharomyces rouxii* occurs. As a result, 2 to 3 percent ethanol and many kinds of aroma components are produced by this yeast. At the same time, phenolic compounds such as 4-ethylguaiacol (4EG) and 4-ethylphenol, which add characteristic aroma to soy sauce, are produced by other types of yeasts such as *Candida versatilis* and *Candida etchellsii*.

It takes over 6 months for the entire fermentation and aging of the *moromi* mash. Therefore, shortening this period is important and new processes for soy sauce brewing are desirable. This paper describes the continuous production of soy sauce in a bioreactor system, which consists of reactors containing immobilized glutaminase and immobilized cells of *P. halophilus*, *Z. rouxii*, and *C. versatilis*.

MANUFACTURING PROCESSES

The processes for soy sauce production using the conventional and bioreactor methods are shown in Figure 1. The bioreactor method differs from the conventional one in the following ways: (a) proteases

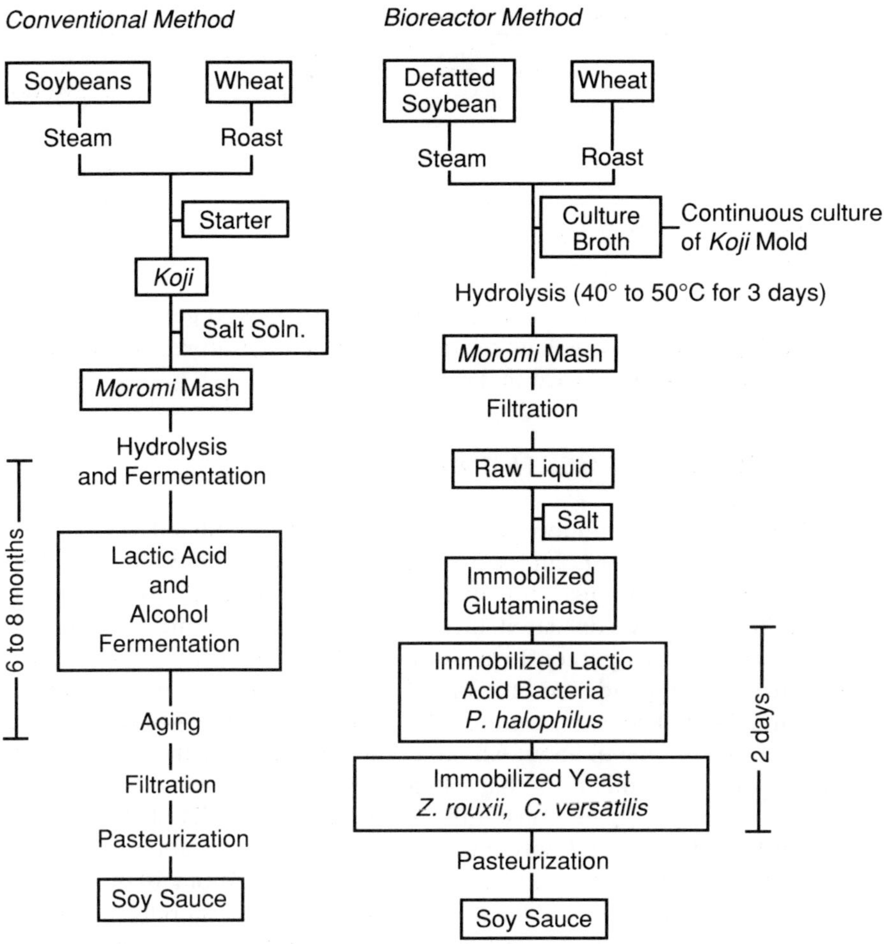

FIGURE 1 Manufacturing processes for soy sauce by conventional and bioreactor methods.

from continuous submerged culture are used (1), (b) fermentation is carried out in the liquid state, and (c) the fermentation period is considerably shorter. It takes several months for the conventional fermentation but only about 2 days for the bioreactor method.

In the bioreactor method, raw liquid was successively passed through, first, a glutaminase reactor to increase glutamic acid; second, a *P. halophilus* reactor to carry out lactic acid fermentation; and, third, a *Z. rouxii* reactor to carry out alcohol fermentation and a *C. versatilis* reactor to produce phenolic compounds such as 4-ethylguaiacol. Two reactors containing immobilized yeast cells were set in parallel, and the flow rate of the feed solution to the *Z. rouxii* and *C. versatilis*

TABLE 1 Conditions of Fermentation in Each Reactor

Reactor	Carrier	Column Volume (L)	Packed Gel Volume (L)	Residence Time, hours	Tempera-ture, °C	Aeration, vvm
Glutaminase	Chitopearl	1.8	0.6	0.7	40	—
P. halophilus	AS	7.5	5.0	6.1	27	—
Z. rouxii	Al	27.0	8.0	25.5	27	0.005
C. versatilis	Al	1.0	0.2	10.7	27	0.08

AS, Alginate-colloidal silica.
Al, Alginate.

reactors was set in a ratio of 10 to 1. Carrier, packed gel volume, and operating conditions such as residence time, temperature, and aeration in each reactor are shown in Table 1.

CONTINUOUS FERMENTATION

A profile of continuous fermentation by immobilized cells of *P. halophilus, Z. rouxii*, and *C. versatilis* is shown in Figure 2. The fermentation continued for over 100 days without any microbial contamination. A consistent increased level of glutamic acid (in the range of 0.3 to 0.4 percent) was found in the effluent from glutaminase reactor, with a residence time of 0.7 hours. Lactic acid was produced by immobilized cells of *P. halophilus* in quantities of 0.7 to 1.0 percent at a residence time of about 6 hours, and consequently the pH declined to 4.9 to 5.0, similar to that of conventionally brewed soy sauce. Ethanol was produced constantly by immobilized cells of *Z. rouxii* in quantities of 2.5 to 2.7 percent at a residence time of about 26 hours. This is the standard ethanol content in soy sauce. About 10 ppm (parts per million) of 4-ethylguaiacol was produced by immobilized cells of *C. versatilis* at a residence time of about 10 hours, and the final 4-ethylguaiacol content after mixing the two fermented liquids from the reactors of *Z. rouxii* and *C. versatilis* was about 1 ppm, which is the optimum concentration in conventional soy sauce. The total residence time for lactic acid and alcohol fermentation was about 30 hours in this system. This was considerably shorter than the conventional fermentation period of 3 to 4 months required to produce the same amounts of lactic acid and ethanol.

High numbers of viable cells were present in the gel and liquid in each reactor. The number was 10- to 100-fold higher in *moromi* mash. The shortening of the fermentation period in the bioreactor method is possibly due to the high density of immobilized cells in the gel and free cells in the liquid.

The main chemical components of the fermented liquid from the bioreactors were examined, including lactic acid, glucose, ethanol, and nitrogenous compounds.

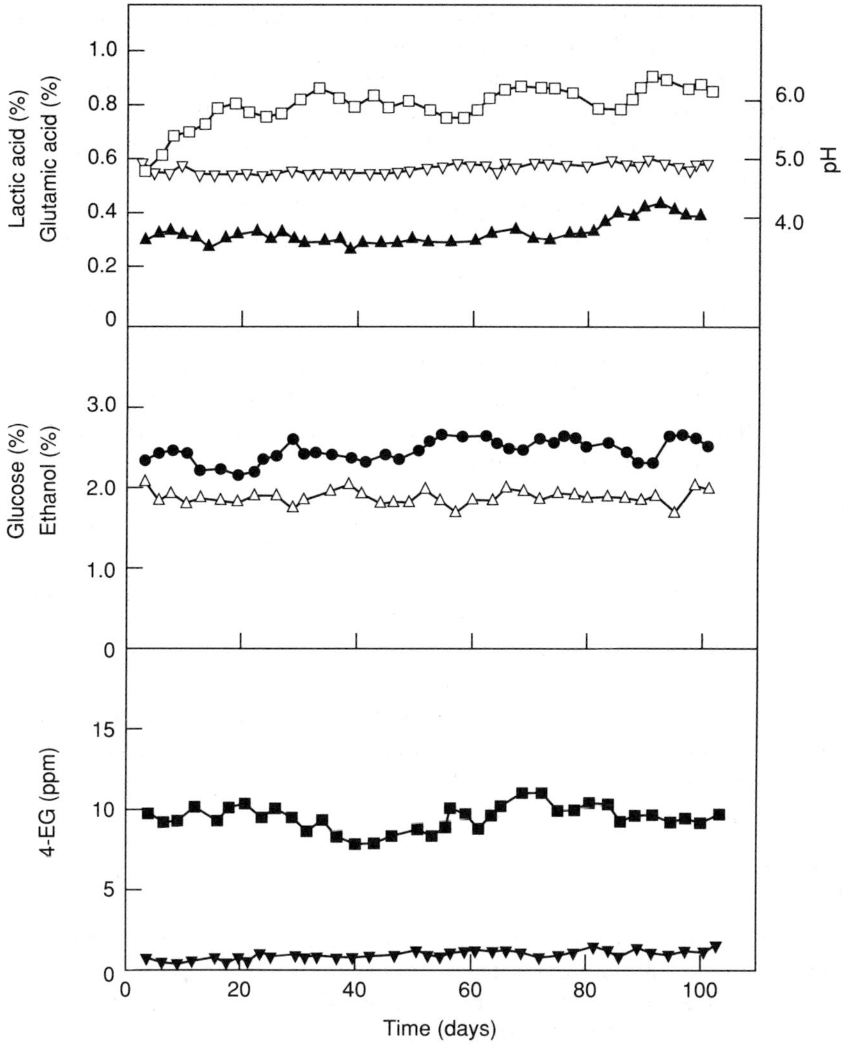

FIGURE 2 Profile of continuous fermentation of soy sauce by a bioreactor system. (□), lactic acid; (▲), glutamic acid (values indicate the increase in the amount of glutamic acid); (▽) pH; (●), ethanol; (△), glucose; (■), 4-EG after passing through the *C. versatilis* reactor; (▼), 4-EG in the final product.

PROPERTIES

The organic acids and aroma components in the bioreactor soy sauce were examined. The proportions of organic acids except citric acid were not much different between the bioreactor soy sauce and the conventional one, although the former was a little lower in acetic acid and succinic acid. It appears that the high residual content of citric acid in the bioreactor soy sauce arises from the inability of *P. halophilus* to utilize citric acid. Aroma components present in both the bioreactor and conventional soy sauces were not qualitatively different. However, the former was higher in isoamyl alcohol and acetoin and lower in isobutyl alcohol, ethyl lactate, 4-hydroxy-2(or5)-ethyl-5(or2)-methyl-3(2H)-furanone, and 4-hydroxy-5-methyl-3(2H)-furanone.

To evaluate the aesthetic qualities of the bioreactor-produced soy sauce, sensory tests were carried out. For example, the intensity of the alcoholic, fresh, sweet, acid, and sharp odors as well as the special *higa* (baking aroma) and *bushoshu* (foul fermented aroma) were compared between the bioreactor and conventional soy sauces. The odors are important for the quality of soy sauce. Although the bioreactor soy sauce was a little weaker in aroma and fresh odor than the conventional soy sauce, the quality of the former was generally judged to be similar to that of the latter.

The total time required for the production of soy sauce by the bioreactor system, including enzymatic hydrolysis of the raw materials, fermentation with immobilized whole cells, and the refining process, is only about 2 weeks (2). This is considerably shorter than the 6 months with the conventional method of soy sauce brewing consisting of *koji* making, fermentation and aging of *moromi*, and refining.

From these results we conclude that the quality of the bioreactor soy sauce was very similar to that of the conventional soy sauce from both chemical and sensory evaluations and that the bioreactor system is practical for the production of soy sauce.

REFERENCES

1. Fukushima, Y., H. Itoh, T. Fukase, and H. Motai. 1989. Applied Microbiology and Biotechnology 30:604–608.

2. Hamada, T., M. Sugishita, Y. Fukushima, and H. Motai. 1991. Process Biochemistry 26:39–45.

V. ANIMAL DERIVATIVES

17

Using Mixed Starter Cultures for Thai *Nham*

Pairote Wiriyacharee

Nham is traditionally made from fresh lean pork that is trimmed; minced; mixed thoroughly with salt, potassium nitrate, cooked rice and seasonings; and packed in either banana leaves (1) or cylindrical plastic bags (2). *Nham* production in Thailand depends on chance contamination with wild organisms—lactic acid bacteria and nitrate-reducing bacteria. It is a long process; generally the fermentation lasts 3 to 5 days depending on the season. When *nham* is packed into cylindrical plastic bags, which exclude air, and is held in the bags during fermentation, a microenvironment is selected for microorganisms that are not only salt tolerant but can also grow in the absence of air. In these gram-positive fermentative types of microorganisms, lactic acid bacteria are predominant (3,4). The fermentable carbohydrates are used by those organisms to produce organic acids, mainly lactic acid, that contribute to a variety of flavors and textures. The *nham* finally develops approximately 1.0 percent total acidity as lactic acid and the pH is 4.3 (5).

MARKETING PROBLEMS

Problems in marketing traditional *nham* include its short shelf life and high price and the intensive labor required for its production. It has high energy costs if kept under refrigeration in the marketplace. Additionally, the manufacturers have a heavy exposure to risk of losing a large stock through a process failure. Pork meat is quite expensive, and the raw material cost is increasing more quickly than the selling price. In addition, large-scale production of *nham* has the problem of its short storage life. A longer shelf life is required so that the *nham* can be distributed to the marketplace. Therefore, the *nham* market needs the product to have consistent quality, safety, and longer shelf life. The *nham* should stay fresh and not turn rancid or develop an off-flavor or change in color when it is in the marketplace.

On the other hand, *nham* production depends on natural fermentation; the product quality therefore varies from batch to batch. The shelf life of *nham* is quite short—approximately a week at Thai ambient temperatures. Chilled conditions can extend the shelf life, but normally the product is stored at ambient temperatures. The sanitation conditions of the processing are also poor because of a lack of knowledge and technology. The initial native lactic acid bacteria may be insufficient to bring about the normal ripening process. This may allow pathogenic bacteria to grow before lactic acid bacteria occur, resulting in the possibility of food poisoning. Since most *nham* is consumed without further cooking, proper fermentation is of paramount importance in ensuring the product's safety.

Somathiti (6) found that the initial coliform count was high in *nham*—approximately 10^7 cells per gram—and decreased to 10^2 cells per gram on the fifth day. An investigation of *Salmonella* in *nham* in the Bangkok market showed that it was present in 56 (or 12 percent) of 450 samples. In *nham* produced in Chiang Mai, Chiang Rai, and Ubonratchathani, *Salmonella* was found in 25 percent, 42 percent, and 11 percent, respectively, of the total samples. However, *Shigella* sp. was not found in *nham* bought from any of these markets.

Thus, the *nham* process needs to be studied to improve product quality, to give a more uniform standard quality, and to develop the technology for applying of the process on an industrial scale before launching extensively in the Thai and export markets.

NHAM DEVELOPMENT

In developing of an improved *nham* process, not only is there a need for the knowledge of modern scientific discoveries and technological developments but also the knowledge of consumers' needs and wishes. The final product must be acceptable to consumers. A unified system is required that combines scientific and consumer information for systematic development of the *nham* product.

Effect of Starter Cultures

In our research mixed starter cultures and the carbon sources used in *nham* formulation were important factors in determining product quality (7). The starter cultures had a potential to make a good *nham* quality. Cooked rice, a carbon source for lactic acid production by starter cultures, was an important factor in *nham* fermentation.

The addition of *L. plantarum* to the *nham* mass accelerated very distinctly the decrease in the pH of *nham*. Consequently, the firmness and color developed, influenced directly by acid production. Those findings were in agreement with the work of many researchers (8-12). *P. cerevisiae* increased the firmness later during the last period of fermentation. The optimum growth of *P. cerevisiae* is at pH 5.0 (13), the conditions during this period allow good growth and acid production causing the increase in firmness. *L. plantarum* inoculation had a very distinct effect in terms of firmness development when it was used together with *P. cerevisiae*.

M. varians in the *nham* system significantly reduced nitrate to nitrite during the initial fermentation and increased the tristimulus values at the beginning of fermentation. *L. plantarum* then continued to intensify the color. This finding agreed with the work of Deibel et al. (14); they reported that nitrate-reducing activity generally occurred during the first 2 to 16 hours, while acid production was initiated after 8 to 16 hours. It was clear that it was important to ensure the nitrate-reducing activity of the *M. varians* that took place prior to its inhibition by the growth of lactic acid bacteria. The nitrite formed was decomposed spontaneously in acid surroundings into nitric oxide, which subsequently reacted with myoglobin to form a pink compound—nitrosomyoglobin. So the residual nitrate in the *nham* system reduced quickly when acid was produced. The rate of nitrosomyoglobin formation increased with falling pH, and this reaction takes place best in the pH range of 5.0 to 5.5 (15) and was therefore accelerated by *L. plantarum*. The *L. plantarum* inoculation had a very distinct effect in terms of color development when it was used together with *M. varians*.

On the other hand, *L. brevis* seemed to be a poor lactic acid producer and decreased the color of the product and also produced gas, which decreased the firmness of the *nham*.

Microbiological Quality

The starter cultures *L. plantarum* and *P. cerevisiae* increased during the initial fermentation and were highest on the third day of fermentation with 10^6 to 10^9 cfu/g^{-1} (colony forming units) and then decreased slowly during the later period of fermentation. In the *nham* sample, on the other hand, the *M. varians* decreased during the fermentation approximately 2 log cycles by the third day. The total bacterial count was related to the starter cultures counts, but there was a little higher count of approximately 1 log cycle. No yeasts or molds were detected in the finished *nham*.

The pathogenic bacteria, including Enterobacteriaceae and *Staphylococcus aureus*, decreased during fermentation. In the *nham* fermented for 3 days the Enterobacteriaceae and *S. aureus* counts were 10^2 and 10^3 cfu/g^{-1}, respectively.

FERMENTATION DEVELOPMENT

In Thailand large amounts of cooked rice are added to the raw *nham* mixture. It is degraded only slowly and may result in growth of undesirable organisms during fermentation, particularly at high ripening temperatures. Glucose is therefore added to the cooked rice. This ensures a sufficiently rapid initial growth and nitrate reduction by *M. varians* and rapid later pH drop, without inhibiting the chemical reactions necessary for the development of firmness and desired color.

Cooked rice and glucose had no effect on pH reduction during *nham* fermentation. As the fermentation time increased, the pH decreased. The pH dropped rapidly after 18 hours of fermentation at 30°C, 43 percent relative humidity with pH 5.1. The beginning of cooked rice reduction coincided with the increase in reducing sugars after 12 hours of fermentation. The reducing sugars declined after another 12 hours of fermentation, and this coincided with the decrease in pH. This indicated that if both cooked rice and glucose were used at high levels (10 percent and 1 percent, respectively) at the beginning of the fermentation, the pH dropped more slowly than if lower levels were used (8 percent cooked rice and 0.5 percent glucose). Increasing the amount of cooked rice, on the other hand, reduced the firmness of the *nham*. There was an increase in weight loss at the high level of glucose. There were 1.0 to 1.3 percent reducing sugars and 2 to 3 percent cooked rice in the finished *nham*, and this residual carbohydrate could be used by the undesirable organisms during storage. Therefore, the carbon source levels in *nham* should be reduced.

When the glucose level was maintained at 0.5 percent but the level of cooked rice increased, a longer period was required to attain adequate fermentation end products (16).

It was also found that 6 percent cooked rice with 0.5 percent glucose in the *nham* formulation, when fermented with starter cultures at 30°C and 97 percent relative humidity, caused rapid pH reduction. Acid production was good, firmness and color development were satisfactory, and the product was microbiologically safe.

The rate of fermentation and the ultimate pH of *nham* are directly influenced not only by the specific formulation but also by the processing conditions. Since the safety and quality of *nham* depend on the rate and extent of acid production, a thorough understanding of these environmental parameters is essential for total control of the product. In our research, higher temperatures increased the rate of fermentation, reduced pH, and improved firmness and color development. The initial temperature of *nham* was very important in determining the final product. The achievement of lowering pH was affected by the initial product temperature and the time at that temperature. For experimentation with frozen meat, the temperature of *nham* mixtures was 15°C; with fresh meat the temperature was 26°C. The pH dropped more quickly in *nham* made with fresh meat than with frozen meat.

Nham made using frozen meat was fermented at 30°C and 97 percent relative humidity. It took 3 days to reduce the pH to 4.3 to 4.4, while the *nham* using fresh meat fermented under the same conditions needed only 2 days to reduce the pH to 4.1.

Nham is usually held at a high temperature during processing to ensure rapid fermentation, but this can also accentuate the growth of pathogens. In addition, *nham* is usually eaten without further cooking by the consumer. These conditions make strict control of the product essential. Although proper sanitation, employee hygiene, and the control of raw materials definitely reduce contamination, ultimate control of product safety must be inherent in the formulation and process. The addition of starter cultures can provide sufficient microbial numbers to ensure numerical dominance over the natural flora, including pathogens, and in combination with the proper processing controls can guarantee the safety and quality of the final *nham*.

Shelf Storage

Nham is usually sold in Thai markets at ambient temperatures (20° to 30°C). It was found that *nham* prepared using the improved conditions described here when stored at these temperatures had a shelf life of 9 to 11 days while commercial *nham* usually has a shelf life of only 3 days. In supermarkets *nham* is stored at chilled temperatures (5°C), and it can be exported at low temperatures (1°C). Additionally, consumers usually store the product in a household refrigerator (10°C). It was found that shelf life was extended to 63 to 103 days at storage temperatures of 1° to 10°C. The higher the storage temperature, the greater the change in *nham* quality.

Sensory Evaluation

Nham fermented with 10^3 cfu/g *M. varians*, 10^3 cfu/g *L. plantarum*, and 10^6 cfu/g *P. cerevisiae* with 6 percent cooked rice and 0.5 percent glucose at 30°C, 97 percent relative humidity for 3 days, was accepted by the trained panel, with an overall acceptability mean ideal ratio score of 0.95±0.01. For quality degradation during storage, the overall acceptability of the product depended on sourness and off-flavor detected in the sample.

Nham using fresh meat fermented at a low temperature was given a higher than ideal score for sourness. However, the newly developed formulation for *nham* was superior to that of the commercial *nham*.

The consumer panel was also used to determine the effect of reducing the fermentation time from 3 days to 2 days. The results showed that only visual texture was significantly different from the ideal product.

In consumer testing the majority of the consumers (90 percent) accepted the developed *nham* in terms of sourness, spiciness, and saltiness.

In conclusion, the development of traditional fermented pork sausage, *nham*, was very successful in that the product was developed by using mixed starter cultures and had a very high quality in terms of consistency, microbiological safety, and longer shelf life. It was also acceptable by the target consumers. The product could be processed in a simple plant and with equipment that was available at the fermented meat factory with only an improvement in the technology of culture preparation and temperature control. In addition, the developed *nham* had a longer shelf life than commercial *nham*. The product, therefore, could be shipped from the cottage industry producers in the north to all provinces in Thailand, particularly to Bangkok, and also gave the potential for overseas shipment if refrigeration is used.

REFERENCES

1. Adams, M. R. 1986. Progress in Industrial Microbiology. Vol. 23. Microorganisms in the Production of Food. New York: Elsevier Science Publishers.

2. Pakrachpan, L. 1981. Fermented Food Industry. (In Thai). Biotechnology Department, Faculty of Agro-Industry, Kasetsart University, Thailand.

3. Comenuanta, J. 1966. Thai Fermented Pork. I. Microbiology of the Thai Fermented Pork. B.Sc. thesis, Kasetsart University, Thailand.

4. Techapinyawat, S. 1975. Microbial Study During Fermentation of Thai Fermented Pork. M.Sc. thesis, Kasetsart University, Thailand.

5. Wiriyacharee, P. 1990. The Systematic Development of a Controlled Fermentation Process Using Mixed Bacterial Starter Cultures

for *Nham*, a Thai Semi-dry Sausage. Ph.D. thesis, Massey University, New Zealand.

6. Somathiti, S. 1982. A Survey of Some Enteropathogenic Bacteria in Thai Fermented Pork. M.Sc. thesis, Kasetsart University, Thailand.

7. Wiriyacharee, P., M. D. Earle, D. J. Brooks, G. Page, and L. Rujanakraikarn. 1991. Identifying the important factors affecting the characteristics of *nham*. Food 21(1):48-58.

8. Klemet, J. T., R. G. Cassens, and O. R. Fennema. 1973. The association of protein solubility with physical properties in a fermented sausage. Journal of Food Science 38:1128-1131.

9. Klemet, J. T., R. G. Cassens, and O. R. Fennema. 1974. The effect of bacterial fermentation on protein solubility in a sausage model system. Journal of Food Science 39:833-835.

10. Klettner, P. G., and W. Rodel. 1978. Testing and controlling parameters important to dry sausage ripening. Fleischwirtschaft 58:57-60, 63-64, 66.

11. Klettner, P. G., and P. A. Baumgartner. 1980. The technology of raw dry sausage manufacture. Food Technology Australia 32:380-384.

12. Palumbo, S. A., L. L. Zaika, J. C. Kissinger, and J. L. Smith. 1976. Microbiology and technology of the pepperoni process. Journal of Food Science 41:12-17.

13. Buchanan, R. E., and N. E. Gibson. 1974. Bergey's Manual of Determinative Bacteriology. Baltimore, Md.: Williams and Wilkins Co.

14. Deibel, R. H., C. F. Niven, and D. D. Wilson. 1961. Microbiology of meat curing. III. Some microbiological and related technological aspects in the manufacture of fermented sausages. Applied Microbiology 9:156-165.

15. Niinivaara, F. P. 1955. The influence of pure bacterial cultures on aging and changes of the red color of dry sausage. Thesis, University of Helsinki, Finland, Acta Agralia Finnica No. 84.

16. Pezacki, W. 1974. Technological control of dry sausage ripening. VIII. Effect of pre-drying on the dynamics of carbohydrate changes taking place at the beginning of ripening. Fleischwirtschaft 58:124-126, 129-132, 135.

18

Starter Cultures in Traditional Fermented Meats

Margy Woodburn

Fermentation traditionally offers an easy and low-energy preservation method for meats that results in distinctive products that have an important part in the diet of people making them. Such fermented meats contribute both nutritional value and pleasure to meals. However, products are not the same from time to time. Indeed, the product may spoil, cause illness due to pathogenic microorganisms or their toxins, and even become lethal due to botulinum toxin production if the normal beneficial microbial flora do not multiply as usual. To prevent these problems, the use of starter cultures has become commonplace in many countries, including developing countries.

One example of such fermented meat is *nham*, a traditional Thai sausage. *Nham* is made by mixing salt (3 percent by weight) and garlic with ground lean pork. Nitrate and nitrite salts also are added in commercial production. The mixture is then wrapped in a banana leaf or stuffed in cellulose tubing. Fermentation is at ambient temperature (about 30°C in Thailand) for 3 to 4 days, after which it remains in good condition for only 1 to 2 days without refrigeration. Since *nham* is frequently eaten raw, it is important that pathogenic bacteria be killed as well as that botulinum toxin and staphylococcal enterotoxins are not produced. Since hogs are frequently infested with *Trichinella spiralis*, these larvae should not be viable.

A study was conducted on *nham* made with and without the addition of one of two levels of a commercially available dry starter culture preparation (Griffith Laboratories, Ltd., Thailand) (1). Portions in polyethylene film bags were inoculated, sealed, and incubated at 30°C. The inoculum was *S. aureus* (a mixture of three enterotoxin-producing strains) and *E. coli* (three strains). Microbial numbers, pH, and titrable acidity were determined at intervals during the fermentation. The meat used was from two hogs that had been experimentally infected with trichinae at weaning; viable trichinae were determined at 24-hour intervals.

S. aureus was able to multiply (10×) and remain viable only in the control inoculated samples. *E. coli* was not detected at 96 hours in the sausage made with the higher level of starter (1.5 percent by weight) and had decreased greatly in products made with the 0.75 percent level. The use of the higher level of starter preparation resulted in loss of infectivity of the trichinae larvae, although further research is necessary to confirm this effect. The addition of starter culture resulted in more rapid acid production and slightly lower end-point pH.

It is important to keep in mind that natural fermentations are difficult to replicate in other settings. For example, the meat mixture for *nham* is traditionally wrapped in small banana leaf packets. The leaves contribute to the surface flora of the sausage, which no doubt changes the fermentation pattern. Flora of work surfaces and of the pork itself may be different.

Drying often follows fermentation of similar meat products to provide for long-term preservation. *Dendeng giling*, Indonesian seasoned beef that has only a traditional short fermentation period before drying, was found to have a lower pH and total gram-negative bacteria, staphylococci, and *E. coli* counts when prepared with a starter culture of *Lactobacillus plantarum* than in the traditional manner. Those with a starter culture dried more rapidly at 50°C and had lower water activities (2).

The effectiveness of lactic acid bacteria in suppressing the multiplication of undesirable microorganisms is largely attributed to the production of organic acid. However, additional factors include the production of bacteriocins and hydrogen peroxide. More general effects include competition for essential nutrients.

To maximize the quality, reproducibility, and safety of the product, strains of bacteria are selected based largely on the qualities of self-stability and viability as used, rapid acid production, and desirable product qualities. As in the starter culture preparation used for *nham*, strains of *Lactobacillus* and *Pedicoccus* are the most common (3,4). The compatibility of strains is important, which includes resistance to or lack of production of bacteriocins. In addition to tolerance to the salt and nitrite levels of the mixture, the culture must be active in the temperature range used for the fermentation. The product must have the expected palatability characteristics. No harmful compounds may be produced. These same attributes can be more efficiently arrived at through the application of the techniques of molecular biology.

The success of traditional fermentations depends on the complex interaction of the food components, the natural flora of the ingredients, and the surfaces in contact with the food, atmosphere, and ambient temperature. Our knowledge of these conditions is still limited for many of the fermented meats. Alaskan outbreaks of botulism from native sea and land mammal products may have increased as plastic bags became the common container and the fermentation rate was speeded by placing the container near the stove (5). Thus, as transitions occur from traditional fermentations to new adaptations, knowledge of the basic processes becomes essential.

REFERENCES

1. Petchsing, U., and M. J. Woodburn. 1990. *Staphylococcus aureus* and *Escherichia coli* in *nham* (Thai-style fermented pork sausage). International Journal of Food Microbiology 10:183–192.

2. Darmadji, P., M. Izumimoto, T. Miyamoto, and K. Katoaka. 1990. Lactic fermentation effects on preservative qualities of *dendeng giling*. Journal of Food Science 55:1523-1527.

3. Smith, J. L., and S. A. Palumbo. 1983. Use of starter cultures in meats. Journal of Food Protection 46:997–1006.

4. Bacus, J. N., and W. L. Brown. 1985. The lactobacilli: Meat products. The pediococci: Meat products. Pp. 55–72, 85–96 in Bacterial. Starter Cultures for Foods. S.E. Gilliand, Ed. Boca Raton, Fla.: CRC Press.

5. Wainwright, R. B., W. L.Heyward, J. P. Middagh, C. H. Hatheway, A. P. Harpster, and T. R. Bender. 1988. Food-borne botulism in Alaska, 1947-1985: Epidemiology and clinical findings. Journal of Infectious Diseases 157:1158-1162.

19

Fermented Fish Products in the Philippines

Minerva SD. Olympia

In many parts of the world especially in Asia, fermented foods are popular and well liked by the general populace and so widely used that the daily diet of the people would not be complete without them. In a developing country like the Philippines, where many fermented food products are known, their popularity is due not only to their characteristic flavor but also to the fact that other processing methods, such as freezing and canning, are generally expensive.

Despite their popularity, research and development on fermented foods is meager. Most of the traditional food fermentation industries especially in the Philippines are rural, seasonal, labor intensive, informal, and capital deficient. Commonly, fermented foods are sold and consumed in the areas where they are produced.

The methods of processing were developed in homes and improvements were based on the observations of the practitioners. Fermentation processes are normally handed down from generation to generation. There is little interest in knowing the role of microorganisms and the physical and chemical changes that occur in the products. What is recognized are changes in color, odor, and taste that result from modifications of the process or variations in the ingredients or conditions. Most processes are conducted on a trial-and-error basis with little quality control. Product quality primarily depends on the experience of the processor.

In the Philippines, fermented fishery products can be divided into two groups. The first group includes those containing high concentrations of salt—about 15 to 20 percent in the final product. This group consists of *bagoong* (fish paste) and *patis* (fish sauce). These products are generally used as condiments.

The second group includes *burong isda* (fermented rice fish mixture) and *burong hipon*, also known as *balao balao* (fermented shrimp rice mixture). These products, when fermented, become acidic with a cheese-like aroma.

FISH PASTE (*BAGOONG*)

Product

Bagoong is the undigested residue of partially hydrolyzed fish or shrimp. It has a salty and slightly cheese-like odor (Figure 1). The characteristics of this product vary depending on the region where it is made and consumed. In the Tagalog provinces, the fish paste is completely fermented and ground, with or without coloring matter added. In the Ilocos region and Pangasinan provinces, the products are either partially or completely fermented. In the Visayas and Mindanao, the product is slightly fermented without liquid; the fish is hard and solid salt is present (1).

Preparation

The fish used for *bagoong* include anchovies, sardines, herring, silverside, shrimp, slipmouth, freshwater porgy, oysters, clams, and other shellfish. The fish are washed thoroughly and drained well. Salt is mixed with the drained samples at varying proportions from 1:3 to 2:7 depending on the bulk of the preparation. The mixture is allowed to ferment for several months or longer until it develops the characteristic flavor and aroma of *bagoong*.

Bagoong is eaten raw or cooked and is generally used as flavoring or condiment in many traditional recipes. As an appetizer it is sauteed with onions and garlic and served with tomatoes or green mangoes. In rural areas, *bagoong* is eaten with vegetables, and, especially in the coastal regions, it is often the main source of protein in the diet.

Microbiological Analysis

Results of earlier studies on the microbiological changes in *bagoong* showed that the total viable count decreased with time. Aerobic organisms predominate at the onset of the fermentation followed by the microaerophilic and anaerobic microorganisms at the later stages (2,3). Information gathered on the microflora indicated that both desirable and hazardous microorganisms are present in this product.

FISH SAUCE (*PATIS*)

Product

In the Philippines the production of fish sauce is always accompanied by the equally important product *bagoong*. This product is the clear supernatant yellow-brown liquid obtained by decanting and/or pressing or centrifuging *bagoong* after it has been thoroughly fermented. Fish sauce may be obtained either from fish or shrimp *bagoong* after 1 to 2 years of fermentation. The longer the digestion period, the better.

Preparation

The raw material used is similar to that of the fish paste. They differ only with respect to the period of fermentation. To obtain the fish sauce, the fermentation is continued until liquid forms on top of the mixture, after which it is drained and filtered.

Microbial Analysis

The total bacterial counts decreased rapidly up to the sixth month and declined slightly until the end of fermentation. Most of the organisms isolated were facultative anaerobes.

Chemical Changes

The solid material is progressively digested with the protein, gradually solubilized by enzyme action, leading to increases in peptides and amino acids in the liquid component. The soluble protein/polypeptide ratio was found to be relatively constant after 1 month. This suggests that most of the proteolytic activities occurred in the early period. Amino nitrogen and total volatile bases (TVBs) increased steadily until the seventeenth day of fermentation. In addition, the lipids in the fish are believed to be broken down during fermentation to yield fatty acids. These may act as precursors for flavor and aroma compounds and may also participate in the browning reactions that increase with prolonged periods of fermentation.

FERMENTED RICE AND SHRIMP (*BALAO BALAO*)

Product

Balao balao is a fermented cooked rice and shrimp (*Penaeus indicus* or *Macrobrachium* species). The mixture becomes acidic during fermentation, and the shrimp shell reddens and softens. It is commonly prepared for the table in sauteed form and is eaten either as an appetizer or main dish.

Preparation

The general method for making *balao balao* is by mixing washed shrimp with salt (about 20 percent w/w) and allowing the mixture to stand for 2 hours or overnight. The shrimp are then drained, mixed with cooled cooked rice, and fermented at room temperature for 7 to 10 days.

Microbial Analysis

The total plate count of this product showed a fluctuating trend. It is believed that this is due to sequencing in the flora involved in the process. Changes in the microflora during fermentation overlap, which suggests that there are changes in conditions during the fermentation that lead to the death of one species and the enhancement of others.

Fermented Fish and Rice (*Burong Isda)*

Product

This product is a popular traditional food in central Luzon. It is usually prepared using freshwater fish. During fermentation the fish flesh becomes very soft and the bones acquire the characteristic softness of cartilage when cooked. Before serving, it is sauteed in oil, garlic, and onion. Similar to *balao balao*, it is consumed either as an appetizer or as a main dish.

Preparation

The method of preparation is almost identical to that for *balao balao*. The fish is scaled, eviscerated, and filleted. It is mixed with salt and allowed to stand overnight before mixing with cooled cooked rice. Fermentation is also carried out for 7 to 10 days at room temperature.

Microbial Analysis

Sequential changes of the bacterial flora also occur in this product and involve the same lactic acid bacterial group as in *balao balao*.

Chemical Changes

During lactic acid fermentation the major chemical change that occurred was the accumulation of lactic acid from the conversion of carbohydrates. This results in changes in the composition and acidity of the product (4). Such changes are attributed to the lactic acid bacteria, which are also referred to as microaerophiles. Changes caused by microaerophiles do not result in the decomposition of the food to its basic components such as CO_2, and H_2O (5). Instead, the most common end product of their metabolism is lactic acid.

Research and Development

At present, the technological know-how for the improvement of traditional fermented fishing products in the Philippines is not advanced. This holds true in the case of the *burong isda* process, which will be described in detail.

Burong isda is a traditional fermented fish product in the Philippines. It is similar to *naresushi* or *funasushi* of Japan. Previously consumed as condiment (6), it is now often a main dish because of economic conditions. *Burong isda* is available in two forms, depending on consumer preferences in a particular area. One is called white *burong isda*, which has a natural product color, and the other is red *burong isda*, which is colored by the addition of *angkak* or *anka. Angkak* or *anka* is a culture of *Monascus purpeveus* grown on rice. The former is preferred in the western provinces of the central Luzon, whereas the latter is preferred in the eastern provinces.

There are several kinds of *burong isda* sold in the market, each named for the kind of fish used. One example is *burong dalag*, a fermented rice-fish mixture using mudfish, *Ophicephalus striatus*. Other kinds are shown in Table 1. Our particular study deals with *burong bangus*, a fermented rice-fish product using milkfish, *Chanos chanos*, or *bangus* in the vernacular. The method of preparation is shown in Figure 1.

TABLE 1 Lactic Fermented Fish Products in the Philippines; Varieties of *Burong Isda*

Name	Local	English	Scientific Name
Burong ayungin	Ayungin	Silver perch	*Therapon plumbeus*
Burong bangus	Bangus	Milkfish	*Chanos chanos*
Burong dalag	Dalag	Mudfish	*Ophicephalus striatus*
Burong gurami	Gurami	Goramy	*Osphronemus goramy*
Burong hito	Hito	Catfish	*Clarias batrachus*
Burong kanduli	Kanduli	Sea catfish	*Arias manillensis*
Burong tilapya	Tilapya	Tilapia	*Tilapia nilotica*
Balao balao	Tagunton	Shrimp	*Macrobrachium* sp.
Burong hipon	Suwahe	Shrimp	*Penaeus indicua*

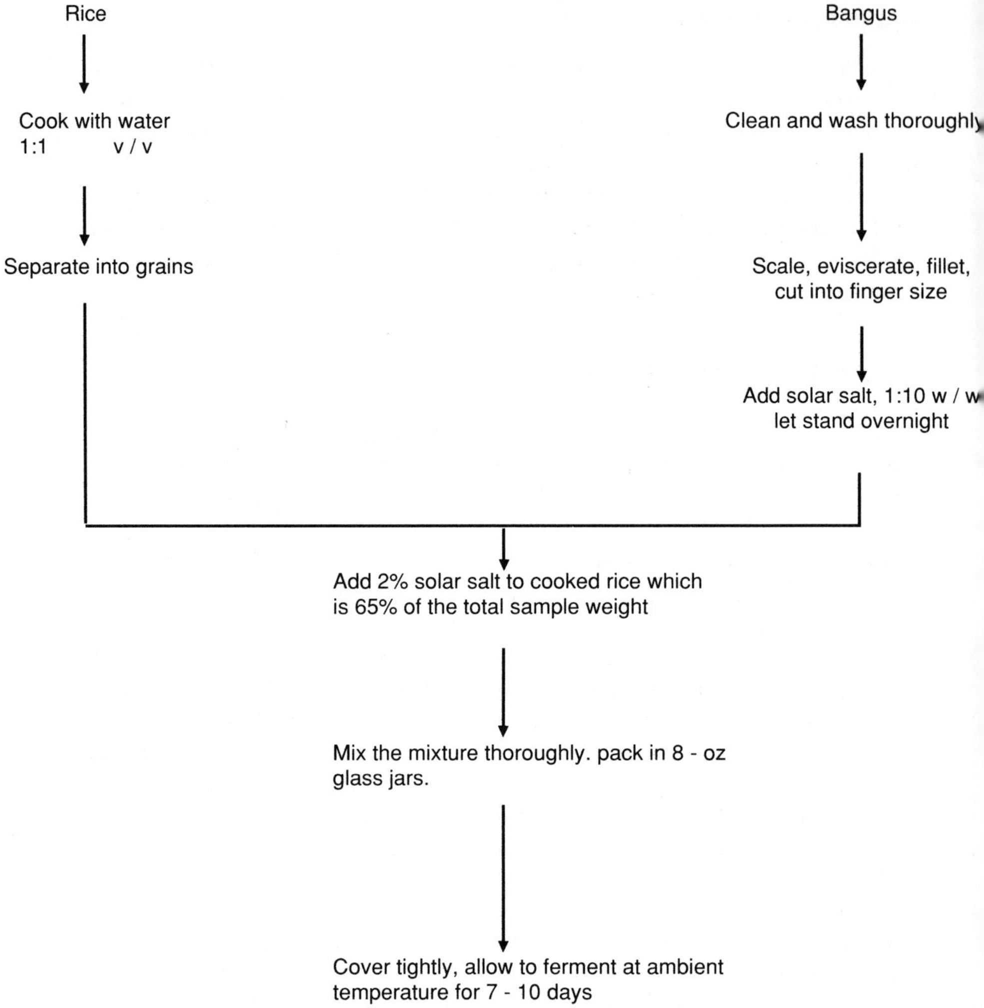

FIGURE 1 The procedure for burong bangus production.

Microbiology

Microbiological analysis of the fermenting mixture showed a sequential type of fermentation with overlapping growth of lactic acid bacteria. The same pattern was also observed by previous workers on other types of *burong isda* (8–10) and in *balao balao* (4, 11). This pattern was not, however, observed in market samples. Instead, only one group of lactic acid bacteria predominated in the product from the onset of the process until it was sold in the market. Market samples were also analyzed for the presence of microorganisms of public health significance. Results showed the presence of coliforms *Salmonella*, and *S. aureus*. The same results were also obtained by Ferrer (12).

ISOLATING STARCH-HYDROLYZING LACTIC ACID BACTERIA

During our study on the microorganisms involved in the fermentation of *burong bangus*, some isolates were found to be capable of hydrolyzing starch (10, 13). Relatively few lactic acid bacteria are known to be starch hydrolyzers, and they were mostly isolated from substrates other than fish. In fact, Bergey's Manual of Determinative Bacteriology (14) does not describe lactobacilli as a starch-hydrolyzing bacteria. However, some of the lactic acid bacteria isolated from *burong bangus* showed otherwise.

The presence of starch-hydrolyzing lactic acid bacteria was also observed in market samples. One of the starch-hydrolyzing isolates, coded L137, was noted to be present in almost all stages of the fermentation process. It was of interest to investigate how the ability of L137 to utilize starch related to its range and level of amylolytic enzyme activity and the type of enzyme(s) that it produces. L137 was characterized and tentatively identified. The enzyme that it produces was also purified. Tentative identification of L137 showed that this strain possesses characteristics similar to *L. plantarum* and *L. cornyformis* subsp. *corneyformis*. L137, however, differs from these two strains in its ability to utilize starch (15). Earlier studies on the lactic acid fermentation of starch-based products reported that lactic acid bacteria cannot hydrolyze starch. The nonacidformers that predominate in the microflora at the onset of the fermentation process, most of which are amylase producers, first hydrolyze the starch to make it available for lactic acid bacteria (16). However, in some *burong bangus* samples no nonacidformers were present; yet fermentation went on. The presence of a starch-hydrolyzing lactic acid bacteria in the fermenting rice-fish mixture, especially during the early stages of the process, will ensure a continuous production of metabolizable sugars for subsequent formation of lactic acid. This will result in a rapid decrease of the pH, thereby inhibiting the growth of other microorganisms that may be amylolytic but that might be possible spoilers and/or human health hazards.

Study of the fermenting samples showed appreciable dextrinizing activity. This would indicate that the sugars formed during the process were mostly oligosaccharides. Our study also showed that acidity increases as the fermentation progresses, even with decreasing reducing sugar production. This would mean that the breakdown products of the starch for lactic acid production were oligosaccharides and reducing sugars. The result also indicates that the lactic acid bacteria involved in the fermentation can utilize oligosaccharides to produce lactic acid. Considering the industrial importance of this strain, the enzyme produced by L137 was purified. Results of the study showed that this enzyme indeed produces oligosaccharides when allowed to react with amylose (15). The activity of this enzyme was found to be highly stable at pH 4 to 5 and is optimum at around pH 4.

REFERENCES

1. Sanchez, P. 1983. Traditional fermented foods of the Philippines. Paper presented at the UNU Workshop on Traditional Food Technologies, July 18–26, Mysore, India.

2. Saisithi, P. 1966. Microbiology and chemistry of fermented fish. Journal of Food Science 31:105–110.

3. Beddows, C. G. 1985. Fermented fish and fish products. Pp. 1–39 in: Microbiology of Fermented Foods, Vol. 2. B. J. Wood (Ed). Elsevier, London.

4. Vatana, P., and R. del Rosario. 1983. Biochemical changes in fermented rice-shrimp mixture. Ph.D. thesis, University of the Philippines at Los Baños, Philippines.

5. Adams, M. R. 1990. Topical aspects of fermented foods. Trends in Food Science and Technology VI(6):140.

6. Barile, E. 1984. Survey of lactic fermented fishery products in central Luzon and their characterization. BSFT thesis, University of Philippines, Diliman.

7. Sakai, H. 1982. The fermented fish food *burong isda* in the Philippines. Journal of Agricultural Science. Tokyo University of Agriculture, Japan.

8. Orillo, C. A., and C. S. Pederson. 1968. Lactic acid bacterial fermentation of *burang dalag*. Applied Microbiology 16(11):1669–1671.

9. Mabesa, A. 1983. Effects of pure culture inoculation on the quality of fermented rich-fish mixture. M. S. food science thesis, University of the Philippines at Los Baños, Philippines.

10. Olympia, M. A., Valenzuela, and M. Takano. 1986. Isolation of an amylolytic lactic acid bacteria in *burong bangus*. Paper presented at the 7th World Food Congress, September 26–October 2, Singapore.

11. Solidum, H. T. 1983. Lactic acid fermentation of *balao balao*. Philippine Journal of Food Science and Technology 7(1):56–86.

12. Ferrer, S. 1981. Microorganisms of public health significance in newly prepared *balao balao*. BSFT thesis, University of the Philippines, Diliman.

13. Olympia, M., and M. Takano. 1991. Lactic acid bacteria in fermented fishery product *burong bangus* (submitted for publication).

14. Rogosa, M. 1974, Gram positive, asporogenous, rod shaped bacteria. Pp. 576–593 in: Bergey's Manual of Determinative Bacteriology. R. Buchanan and N. E. Gibson (Eds.). Baltimore, Md.: Williams and Wilkins Co.

15. Olympia, M., Y. Nakata, T. Date, H. Nakamura, C. Wongkhalaung, A. Shinmyo, and M. Tokano. 1991. Purification and characterization of amylose from a lactic acid bacterium isolated from *burong bangus* (submitted for publication).

16. Yoshioka, F., and M. Kozaki. 1983. Pp. 58–86 in: Summary of Annual Meeting of Nippon Nogei Kogaku Kai (In Japanese).

17. Fleet, G. H. 1986. Biotechnology and the food industry. Food Technology in Australia 3(11):464.

20

Fish-Meat Sausage

Sam Angel and Eliana Mora P.

Fermentation allows the preservation of foods of vegetable or animal origin so that they can be stored and shipped at ambient temperatures and used without further preparation. Lowering the pH is the ideal way to process foods for use in less-developed regions of the world. Proteins, which are needed for growth and development, especially in children, are often in short supply in famine-ravished areas and poor underdeveloped countries. In many areas of the world there are frequently underutilized sources of muscle proteins that could provide excellent starting materials for preservation by fermentation or acidulation.

In Germany a popular noncooked fermented sausage (*rohwurst*) has been produced from beef and pork for many years (1). Lactic acid produced by fermentation lowers the pH of the meat to its gel point, which causes it to firm (2). Further drying increases firmness and reduces water activity. The low pH prevents the development of pathogenic bacteria (3,4), and lower water activity prevents microbial growth and spoilage (5).

The pH can also be lowered by using glucono-delta-lactone, which produces gluconic acid upon contact with water, or using citric or lactic acids. Encapsulated acids release acid more slowly and prevent texture breakdown. In the encapsulation process solid acid granules are coated with hydrogenated vegetable oils or diglycerides, which require heat to release the acid. Graves (6) patented a new water-soluble low-temperature release coating for citric acid. The use of acids directly saves fermentation time, and myofibrillar protein gelation can take place within hours after mixing the meat with acids.

RESULTS AND DISCUSSION

Rohwurst beef-pork sausage served as a model for the development of a similar product from underutilized fish, meat trim, and poultry. A sausage-type product allows the combination of muscle from various sources. The object was to use underutilized muscle protein sources, especially fish, to produce a nutritious and acceptable dry sausage. The product was to be eaten out of hand and thus help to alleviate protein deficiency, especially in children.

Underutilized or inexpensive fish, fish tissue residue from filleting operations, red meat trim, and spent layer hens were the raw materials used in Israel, the United States, and Costa Rica to produce fermented dry sausages.

Cod or haddock frames (i.e., skeletons with residual tissue) were mechanically deboned, and the flesh mince was mixed in equal proportions with either beef or pork trim or mince from mechanically deboned spent layer hens. The batters were mixed with salt, sugar, spices, nitrite, and *Lactobacillus* or *Pediococcus* starter cultures and stuffed into 20-mm collagen casings. They were fermented at 22°C for up to 24 hours depending on pH development.

Control sausages consisted of beef and pork only. The pH of the fish-meat sausages fell to 5.1 to 5.4, while the pH of the beef-pork controls fell to 5.0 to 5.1. Drying took 1 week, at which time the fish-meat sausages contained 17 to 30 percent moisture and the beef-pork controls 25 to 30 percent. Fat content was 17 percent in all the batters at the outset. After drying it was 21 to 30 percent for the fish-meat sausages and 29 to 30 percent for the beef-pork. The fat contents for the fish-meat sausages were significantly lower than for similar commercial sausages in Germany.

Three panel sessions were held. Between 25 and 70 persons participated in each session. All the sausages were found acceptable, as shown in Figure 1. A minority of the participants commented on a fishy taste, especially for the fish-chicken sausages.

In a 3-year cooperative project (7) the flesh of pond-raised silver carp and sea fish in Israel and Costa Rica was deboned and washed. It was then used to prepare fermented or acidulated dry sausages with pork or beef trim or whole-muscle turkey bottom meat. All-fish sausages and 25 to 50 percent fish-meat sausages were prepared. Fermentation was induced with *Pediococcus* plus *Lactobacillus* starter cultures. Acidulation was carried out by adding encapsulated low-temperature-release citric or higher-temperature-release lactic acids.

The pH usually fell to 4.85, except for the fish-turkey sausages where the pH did not fall below 5.0. Starting and final pHs were similar for the fermented and the acidulated sausages, but the pHs for the acidulated sausages fell to their final level within a few hours as compared to several times that for the fermented sausages. Thus, the acidulated blend had a head start on drying. The entire process of pH reduction, firmness development, and subsequent drying was shortened considerably for the acidulated fish-meat sausages.

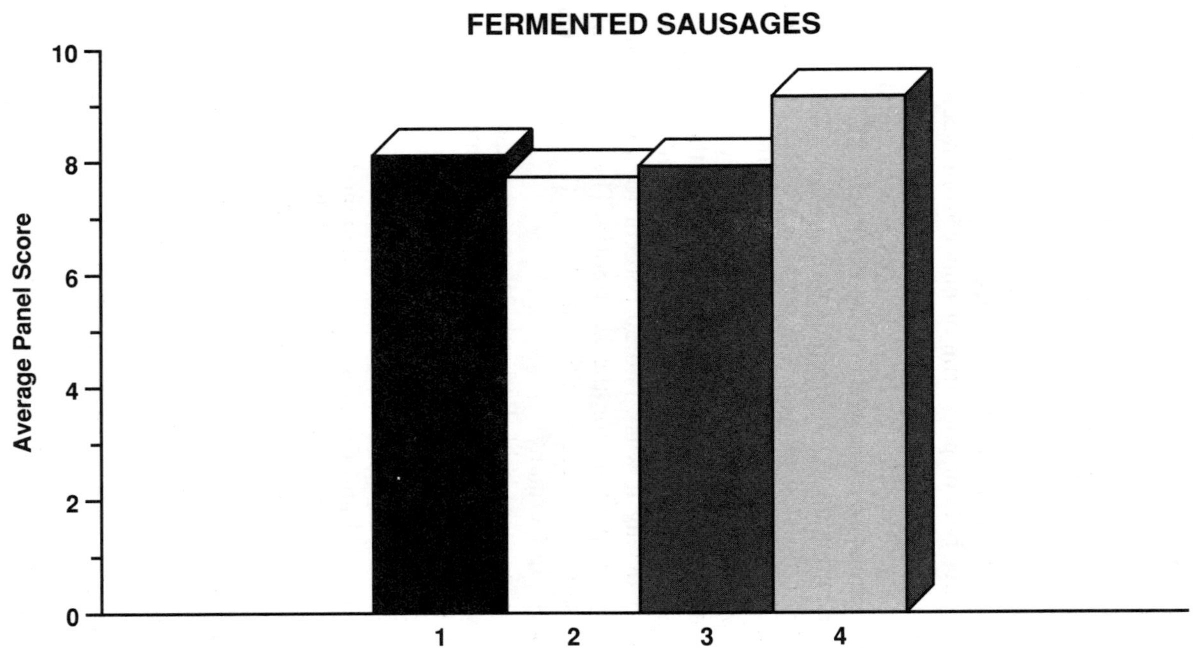

FIGURE 1 Average scores for acceptability of fermented fish-meat and beef-pork sausages. Average scores of three consumer panels in the United States for fermented sausages made from fish-pork, fish-chicken, fish-beef, and beef-pork. Score of 10 means highly acceptable; 5, not acceptable but not rejected; below 5, various degrees of rejection. Fish-pork sausage received the highest average score, 8.1, of the fish-meat sausages. Next highest average score of 7.9 was for the fish-beef sausages. Fish-chicken sausages received a score of 7.7, and nonfish beef-pork sausages received an average score of 9.

Citric acid was found to reduce the pH at lower concentrations than lactic acid, but in 50 percent fish-meat sausages with 0.65 percent citric acid, which was the maximum concentration used, the pH could not be lowered sufficiently. Lactic acid could be used at higher concentrations to lower the pH when necessary. A pH of 4.80 to 4.85 helped the drying process, and lactic acid was of greater benefit in this respect than citric acid.

The experimental sausages were evaluated on a kibbutz in Israel and in 110 households in Costa Rica. The results of the evaluations in both countries were encouraging. In Israel the scores were 5 to 5.6 out of a maximum of 9 for the 50 percent fish sausage. Over 80 percent of the tasters in Costa Rica gave positive responses to the sausages that contained fish (Figure 2). The highest social class was least enthusiastic about the sausages. In Costa Rica the population is not accustomed to eating nonheated sausages. The evaluators therefore either cooked or fried them before eating. The organoleptic tests are consequently being repeated with new instructions.

RECOMMENDATIONS

To minimize production costs, these sausages should contain a minimum of 50 percent fish (from frames or other underutilized sources such as by-catch and trash fish).

Acidulation produced sausage with a good texture, and it can be recommended as a procedure to reduce processing time.

To improve acceptability and nutritional value as well as reduce costs and ensure quality, more research needs to be done on:

- flavor formulation and colorants to meet local population preferences;
- reduction of fat content and introduction of new sources of fat;
- inclusion of soy or other vegetable proteins;
- chemistry and histochemistry of the acidulation and drying processes to improve the efficiency of these steps;
- the effect of replacing nitrite on the wholesomeness of the product [according to Leistner (8), spores of bacilli and clostridia do not grow when there is a sufficiently low pH and low water activity]; and
- protein efficiency feeding for young children and adolescents.

These products should undergo taste tests for acceptability in other Latin American countries as well as other areas with low protein intake.

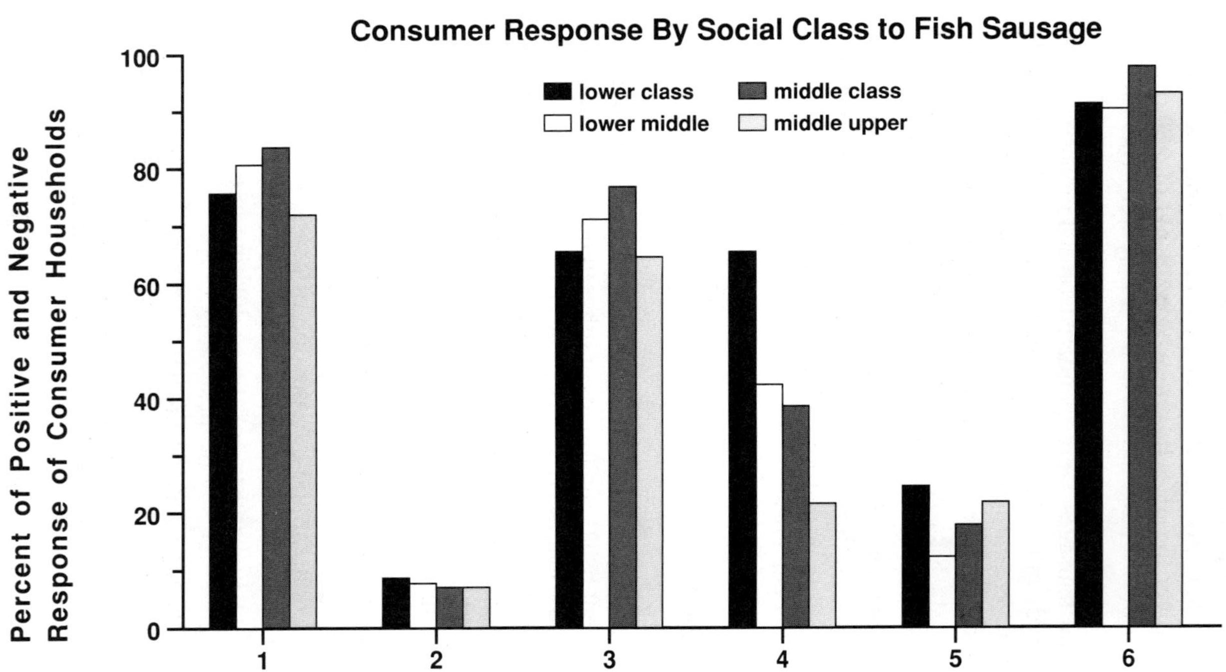

FIGURE 2 Social class of consumers who rated the organoleptic parameters of the fish sausages. Shown are the percent positive and negative and net positive responses to the fish-containing sausages of over 400 persons in 110 households in Costa Rica. Also given are comparisons by social class to other sausages on the market.

REFERENCES

1. Klettner, P. G., and P. A. Baumgartner. 1980. The technology of raw dry sausage manufacture. Food Technology in Australia 32:380.

2. Rodel, W., K. Krispien, and L. Leistner. 1979. Measuring water activity of meat and meat products. Fleischwirtschaft 59:649.

3. Baird-Parker, J. and B. Freame. 1967. Combined effect of water activity, pH and temperature on growth of *Clostridium botulinum* from spore and vegetative cell inocula. Journal of Applied Bacteriology 30:420.

4. Collins-Thompson, D. L., B. Krusky, and W. R. Usborne. 1984. The effect of nitrite on the growth of pathogens during the manufacture of dry and semi-dry sausages. Canadian Institute of Food Science and Technology Journal 17:102.

5. Labots, H. 1981. Aw und pH-wert konzept für die enteilung von fleischerzengnissen in verberbliche und lagerfahige produkte. Fleischwirtschaft 61:1.

6. Graves, R. 1988. Sausage fermentation: New ways to control acidulation of meat. The National Provisioner.

7. Angel, S., and E. P. Mora. 1991. The development of shelf stable fish, poultry and other meat products through energy saving fermentative processes. Final report, Project DPE 55446536035, submitted to USAID, Washington, D.C.-U.S. Cooperative Development Research Program Between Israel and Costa Rica. 62 pp.

8. Leistner, L. 1986. Personal communication.

21

An Accelerated Process for Fish Sauce (*Patis*) Production

R. C. Mabesa, E. V. Carpio, and L. B. Mabesa

The single, probably most important, limitation in the manufacture of fish sauce is the length of time required for its production. It normally takes approximately 12 months from salting to maturity. This limits the turnover rate and overall profitability of a potentially lucrative enterprise. Considering the capital outlay and operating expense required to run a fish sauce business, it is imperative to develop a simple, economical, practicable accelerated process that yields acceptable fish sauce.

With this goal, research and development efforts were undertaken at the food pilot plant of the Institute of Food Science and Technology, University of the Philippines at Los Baños.

OBSERVATIONS

This investigation stemmed from the observation in commercial tanks that freshly drawn fish sauce lacks the desirable aroma of mature sauce; this aroma develops after overnight storage or longer. The appropriate color is there initially but typical flavor is lacking. It was also observed that flavor, aroma, and color development of *patis* in both concrete and wooden vats is more rapid and pronounced during the hot summer months. Constant agitation through pumping and frequent transfer of fish sauce from one container to another also hastened and enhanced development of flavor and aroma. It was hypothesized, therefore, that artificial agitation and/or aeration and heat may help with the development of desirable qualities in fish sauce. Thus, small-scale laboratory experiments were carried out initially. It was determined that timing is of primary importance in the application of heat and aeration. The typical fish sauce characteristics did not develop when freshly salted fish was aerated and heated immediately after mixing. Trials were carried out to determine the appropriate time for aeration and heating of the fish salt mixture. It was found that aging for about a month after salting was sufficient and that higher temperatures resulted in more rapid and greater improvement in quality. However, preliminary experiments indicate that the maximum temperature should not exceed 50°C or a cooked flavor results.

A concrete tank simulating the dimensions of a commercial tank was constructed to test the findings in the laboratory. Technical specifications are given below (see box). It was concluded that fish sauce comparable to traditionally manufactured sauce can be obtained in about 2 months or less using modified reaction conditions. These conditions are given under B and C. Sauce characteristics are given under D.

It is likely that production time may be further reduced if strongly halophilic, proteolytic, and thermophilic *Bacillus* and *Pediococcus* species used in the laboratory can be used in production.

DISCUSSION

Fish sauce with the desirable qualities of traditionally produced sauce was obtained in the pilot plant. The improved process resulted in an acceptable product in about 2 months instead of the 10 to 12 months required for the traditional process. Clearly, savings in time and an improved turnover rate can result if these results are applied commercially. This means greater income-generating capacity.

Some problems, such as loss of volume and contamination with molds and bacteria, were encountered during heating and aeration. The former was remedied by day-to-day monitoring of fish sauce levels and replenishment with plain tap water when necessary. The second problem was resolved by installing cotton filters at the intake end of the pumps and by adding sorbic acid to the sauce at 0.05 percent prior to bottling.

CONCLUSION

With pilot-level success, there is reason to believe that the process can be applied on a commercial scale. However, there are problems attendant to adaptation of the process. Additional expense will be incurred in equipment acquisition, installation, and operation. Heating and aeration alone will increase the price of *patis* by about P 50 per drum or about P O.25 per liter. These costs must be weighed against

TECHNICAL SPECIFICATIONS

A. Tank
   1. Type—concrete, cube approximately 0.265 m × 0.265 m × 0.265m I.D.
   2. Material—concrete 3:2:1 mixture of sand, gravel, and cement with Sahara water proofing added.
   3. Heaters—two 1,000-watt rod-type heaters located close to the center of the tank.
   4. Air pump—one aquarium-type air pump with discharge capacity of 5 liters/minute; pump discharge located 2.5 centimeters below heaters.

B. Operating Information
   1. Preliminary incubation—50 days at ambient temperature.
   2. Air pump—operated 4 hours a day for 10 days.
   3. Heaters—operated 4 hours a day for 10 days.
   4. Temperature—45° to 60°C for 10 days.
   5. Power requirement—7 amperes (pump and heater).
   6. Voltage requirement—220 volts.

C. Raw Material Information
   1. Total weight of fish salt mixture—320 kilograms.
   2. Proportion—1 salt:2 fish by weight (106.6 kilograms salt:213.3 kilograms fish).
   3. Fish species—*Decapterus macrosoma*.
   4. Source—Navotas Fishery Port.

D. Sauce Characteristics
   1. Color—golden yellow-brown highly typical of fish sauce and clear.
   2. Odor—slightly acidic and fishy, typical of fish sauce.
   3. Flavor—typical fish sauce.
   4. Total solids—41 percent.
   5. Protein—14 percent.
   6. pH—6.0.
   7. Salt—24 percent.
   8. Specific gravity—1.21.
   9. Yield—137.5 kilograms.

savings or advantages such as faster turnover rate, decreased overhead, salaries, and power.

Each manufacturer or potential user of a new technology such as this stands to gain substantially despite the additional costs. However, each interested user may find his or her situation unique. A careful study of all terms, factors, and conditions affecting a user should be undertaken before embarking on a new and innovative process such as this.

In light of these results and consequent problems, efforts are under way in the laboratory to reduce process costs, particularly with respect to reducing heating time, minimizing heat losses, increasing heating efficiency, and exploring alternative sources of energy for use in the process.

VI. HUMAN HEALTH, SAFETY, AND NUTRITION

22

Nutrition and Safety Considerations

O. Paredes López

Fermentation was one of the first methods used by Man to produce and preserve foods. Microbial fermentations have played an important role in food processing for thousands of years. Fermentations provide a way to preserve food products, to enhance nutritive value, to destroy undesirable factors, to make a safer product, to improve the appearance and taste of some foods, to salvage material otherwise not usable for human consumption, and to reduce the energy required for cooking (1). Preservation of foods by salting is an age-old practice; while preventing the growth of pathogenic microbes, it allows the development of harmless, halotolerant ones that produce desirable sensorial changes in the substrate (2).

Traditional fermented foods may be divided into two broad categories: (a) submerged culture-fermentations (SCFs) and (b) solid-substrate fermentations (SSFs). In SCFs microbial activity occurs at a relatively low biomass concentration in the liquid phase, while in SCFs microbial growth and product formation occur on the surfaces of solid substrates (3,4). Some examples of traditional fermented foods for SCFs are *pulque* and *tesguino,* soy sauce, fish sauce, kaffir beer, and palm and rice wines. Examples of SSFs are *tempe, miso, pozol, oncom*, and *natto*. One of the major characteristics that distinguishes SSFs from SCFs is that SSF processes usually occur at low-moisture contents (e.g., 10 to 20 percent), conditions under which water activity favors the development of filamentous fungi. However, for many indigenous fermentations the microbial interactions are complex and mixed fungal-bacterial, fungal-yeast, and yeast-bacterial combinations occur (5). These interactions play an important role in the nutritional, safety, and sensory characteristics of the end product (6).

EFFECT OF FERMENTATION ON NUTRITIONAL COMPOSITION

Changes in Proximate Composition and Soluble Components

During fermentation the microorganisms secrete hydrolytic enzymes into the substrate and assimilate some of the fatty acids, amino acids, and simple sugars thus liberated. These are converted into microbial structural components and secondary metabolites. Lactic acid fermentation is an ancient process whereby a varied group of bacteria ferment carbohydrates, producing lactic acid as the major end product. This type of fermentation is used for the production of dairy products, sauerkraut, bread, meat, and silage. In particular, traditional SSFs of legumes, cereals, and starchy substrates have been associated in many regions of the world with the activity of lactic acid bacteria (7); during fermentation lactic acid accumulates, with a concomitant increase in acidity and a decrease of dry matter yields. The higher pH values of fermented legumes, compared to other materials under similar conditions, have been attributed to their higher protein content (8,9).

It seems that the only fermented food showing significant changes in its crude composition is *pozol*. The fermentation mixture contains *Agrobacterium azotophilum*, which is capable of fixing nitrogen (10). Due to the crude methods of analysis, the proximate composition of foods does not change much during fermentation. However, there is almost always a high increase in the soluble fraction of a food during fermentation. The proteolytic activity of bacteria in traditional fermentations degrades complex proteins into simpler proteins, peptides, and amino acids. The bacteria used in *natto* fermentation cause substantial increases in the level of free amino acids and soluble carbohydrates. On the other hand, *Rhizopus* spp., used in the fermentation of various types of *tempe*, are highly hydrolytic, and outstanding increases in soluble fat, protein, and carbohydrate are observed. Free fatty acids, including the essential fatty acids, linoleic and linolenic acids, may increase in these indigenous fermented foods (11,12); this increase is thought to be of nutritional significance.

The increase in soluble solids is a nutritionally desirable event, as the food is effectively digested prior to consumption. In some cases the microorganisms are capable of producing pectinases and cellulases, softening the texture of the food and liberating sugars that would otherwise be unavailable to the human digestive system. Consequently, fermented foods are expected to be more digestible than their unfermented counterparts.

Changes in Composition of Amino Acids and Vitamins

Methionine, the limiting amino acid in legumes, has been reported to increase during *tempe kedele* production, and lysine, the limiting amino acid in cereals, increases during fermentation with *Rhizopus* spp. (1). During *kocho* production, an acidic fermentation, the essential amino acid content is considerably enhanced. On the other hand, during *tape' ketan* and *enjera* production, the levels of some essential amino acids fall, whereas others remain unchanged (11). In general, most traditional fermented foods exhibit slight changes in essential amino acids.

Interestingly, isolation of improved strains of *Aspergillus niger* for an SSF process allowed 200 to 300 percent lysine overproduction compared to the parent strain (5). However, it should be emphasized that bioavailability and balance of amino acids are more important than their total content. Hence, biological experiments to assess their nutritional value are warranted.

Traditional fermentations dramatically improve the vitamin content of a wide variety of substrates. Of all the foods investigated, only *enjera* showed a decline in vitamin content (1,13).

Changes in Unwanted Components

Unwanted components, such as phytic acid, trypsin inhibitor, flatus factors, and lectins, may be present in high concentrations in several desirable foods. Phytic acid and trypsin inhibitor interfere with digestion by binding enzymes. Phytic acid may also bind minerals, reducing their bioavailability. Lectins are capable of binding to the intestinal wall and thus interfering with nutrient absorption. Presoaking and cooking of foods can reduce the levels of some, but not all, of these antinutritional factors. However, microorganisms have the capacity to hydrolyze them, reducing their levels even further (14). Hence, bacteria, yeasts, and fungi that degrade antinutrients at a fast rate and at early stages of fermentation need to be identified or developed (1).

Changes in Biological Value

Since fermentation increases the quantity of soluble proteins in foods, it may improve the amino acid profile, and because it reduces the levels of certain antinutritional factors that interfere with digestion, it would not be unreasonable to suggest that fermented foods will be more efficiently utilized by the human digestive system. Single- as well as mixed-culture fermentations of pearl millet by yeasts improve starch and protein digestibility (15). *Enjera* is one of the few traditional fermented foods that shows a decline in protein efficiency ratio (PER), probably due to a decline in the essential amino acid content (16). Also, increases in PER values of some indigenous fermented foods can be obtained by incorporating soybeans into cereal-based substrates.

SAFETY ASPECTS OF TRADITIONAL FERMENTED FOODS

Because many fermented foods are produced using fungi, the risk of mycotoxin contamination is high. During natural fermentations, food-poisoning flora and coliforms may also grow with the lactics. These microorganisms need to be eliminated to make fermented foods safe for consumption (16). Several factors contribute to the safety of fermented foods: (a) Soaking and cooking. Washing, soaking, and cooking treatments reduce the in situ microbial contaminants. (b) Salting. Various fermented foods are made with the addition of salt, which acts as a preservative. (c) Acid formation. Many indigenous fermentations are carried out by acid-producing microorganisms, where these organic acids (e.g., lactic, acetic, fumaric acids) act as preservatives or as bacteriostatic agents. An inhibitory pH for bacterial growth is considered to be 3.6 to 4.1. (d) Antibiotic production. Molds used in some traditional fermentations produce antimicrobial glycopeptides. (e) Low moisture content. In the case of SSF processes, the low water activity may be an important preservative factor. and (f) Reduction of aflatoxin by some microorganisms. *Rhizopus* and *Neurospora* species, among others, are reported to decrease aflatoxin content of contaminated substrates.

Despite these factors, it has been reported that the sanitary quality of some Oriental fermented foods is poor (17,18). Safe products are usually obtained when the following recommendations are observed: (a) appropriate soaking of the beans in acid at a low pH; (b) adequate cooking time; (c) using hygienic conditions during production, handling, and storage; and (d) good refrigeration of products (5°C) between production and consumption.

In summary, production of foods with high nutritional and sensory values, and free of microbiological health risks, is a key component of any policy aimed at upgrading the social role of traditional fermented foods in less developed countries.

REFERENCES

1. Paredes-López, O., and G. I. Harry. 1988. Food biotechnology review: Traditional solid-state fermentations of plant raw materials. Application, nutritional significance, and future prospects. Critical Reviews in Food Science and Nutrition 27:159–187.

2. Beuchat, L. R. 1978. Traditional fermented food products. Pp. 224–253 in: Food and Beverage Microbiology, L. R. Beuchat (Ed.), Westport, Conn.: The AVI Publishing Co.

3. Tengerdy, R. P. 1985. Solid substrate fermentation. Trends in Biotechnology 3:96–99.

4. Paredes-López, O., and A. Alpuche-Solis. 1991. Solid substrate fermentation. A biotechnological approach to bioconversion of wastes. Pp. 117–145 in: Bioconversion of Waste Materials to Industrial Products, Vol. 1, A. M. Martin (Ed.), London: Elsevier, Applied Science Publication.

5. Rogers, P. L. 1989. Principles and applications of bioprocess technology in the food industry. Pp. 223–239 in: Biotechnology and the Food Industry, P. L. Rogers and G. H. Fleet (Eds.). New York: Gordon and Breach, Science Publishers.

6. Hall, R. J. 1989. Application of biotechnology to traditional fermentations. Pp. 241–277 in: Biotechnology and the Food Industry. P. L. Rogers and G. H. Fleet (Eds.). New York: Gordon and Breach Science Publishers.

7. Fukushima, D. 1985. Fermented vegetable protein and related foods of Japan and China. Food Reviews International 1:149–209.

8. Zamora, A., and M. L. Fields. 1979. Nutritive quality of fermented cowpeas and chickpeas. Journal of Food Science 44:234–237.

9. Paredes-López, O., J. Gonzalez-Castaneda, and A. Carabez-Trejo. 1991. Influence of solid substrate fermentation on the chemical composition of chickpea. Journal of Fermentation and Bioengineering 71:58–62.

10. Cravioto, O. R., Y. O. Cravioto, G. Massiew, and J. Guzman. 1955. El pozol, forma indigena de consumir el maiz en el sureste de Mexico y su aporte de nutrientes a la dieta. Ciencia (Mexico) 15:27–30.

11. Steinkraus, K. H. 1983. Indonesian *tempe* and related fermentations. Pp. 217–251 in: Handbook of Indigenous Fermented Foods, Microbiology Series, Vol. 9, K. H. Steinkraus (Ed.). New York: Marcel Dekker.

12. Paredes-López, O., G. I. Harry, and R. Montes-Rivera. 1987. Development of a fermentation procedure to produce a *tempe*-related food using common beans as substrate. Biotechnology Letters 9:333–333.

13. Soni, S. K., and D. K. Sandhu. 1989. Fermentation of *idli:* Effects of changes in raw materials and physical-chemical conditions. Journal of Cereal Science 10:227–238.

14. Mital, B. K., and S. K. Garga. 1990. *Tempe*—Technology and food value. Food Reviews International 6:213–224.

15. Khetarpual, N., and B. M. Chauhan. 1990. Fermentation of pearl millet flour with yeasts and lactobacilli: In vitro digestibility and utilization of fermented flour for weaning mixtures. Plant Foods and Human Nutrition 40:167-173.

16. Wang, H. L., and C. W. Hesseltine. 1981. Use of microbial cultures: Legume and cereal products. Food Technology 33(1):79–83.

17. Tanaka, N., S. K. Kovats, J. A. Guggisberg, L. M. Meske, and M. P. Doyle. Evaluation of the microbiological safety of *tempe* made from unacidified soybeans. Journal of Food Protection 48:438–441.

18. Samson, R. A., J. A. van Kooij, and E. de Boer. 1987. Microbiological quality of commercial *tempe* in the Netherlands. Journal of Food Protection 50:92-94.

23

Mycotoxic Flora of Some Indigenous Fermented Foods

Felixtina E. Jonsyn

Fermented foods have a wide usage in Sierra Leone as baby/weaning foods. *Ogi* (fermented maize/sorghum) and *foofoo pap* (fermented cassava) are examples. *Foofoo* is also one of the two staples of the Creoles that is now widely used by other tribes especially when rice is scarce. *Ogiri* (fermented sesame seeds) is a favorite condiment used mostly by the poor as a low-cost protein substitute. Several studies (1–4) have shown that toxigenic fungi do not participate in the fermentation processes but contaminate the product during or after the fermentation.

It has been demonstrated (1–4) that at times the substrate for fermentation (maize, sesame seeds) has had prior exposure to mycotoxin. In the case of maize, an aflatoxin B_1 level of 200 μg/kg was reduced to 58 μg/kg in the resulting fermented *mashogi* (5). The long cooking period (6 hours) of sesame seeds before fermentation accounts for the loss of mycotoxins. Studies carried out by Ogunsanwo et al. (6) have shown that losses of 64 percent aflatoxin B_1 and 83 percent aflatoxin G_1 could be observed in *ogiri* product prepared from *Aspergillus flavus*-contaminated melon seeds.

In Sierra Leone, *ogiri* is produced by moist solid fermentation of sesame seeds, a process similar to Nigerian *ogiri*, which is made from fermented melon seeds (*Citrullus vulgaris*) (7) and *Dawa-dawa* from fermented locust beans (*Parkia filicoidea*) (8). Traditionally, the boiled seeds are wrapped in jute bags and allowed to ferment for 4 to 5 days before smoke treatment is applied. In such processes whitish threads are observed after day 2 and molds become obvious after 3 to 6 days

This study was funded by the International Foundation for Science, Stockholm, Sweden.

(3). Microscopic examination of these whitish threads revealed the presence of toxigenic and nontoxigenic *Aspergilli* and *Penicillia* species. Detection of the corresponding mycotoxins of these toxigenic fungi in the fermented, marketed, and stored *ogiri* (4) led to the present study to design appropriate fermentation and storage techniques to reduce the risk of mycotoxin contamination.

MATERIALS AND METHODS

Fermentation Process

Sesame seeds were soaked overnight and pounded in a mortar to dehull. The seeds were then washed and boiled for 6 hours. The boiled seeds were divided into three portions. One portion was transferred to a clean dry nylon fiber bag; the other was placed in a clean dry jute bag. Both were tightly wrapped. The third was placed in a plastic bowl with a tight-sealed lid. Three replicates of each of the nylon fiber and jute bag arrays were made. These were divided into three groups. Group one was left to ferment for 5 days without smoke treatment. Group two received early smoke treatment, from day 2 until day 5. Group three was smoked consistently from day 3 to day 8, and thereafter on alternate days for 2 weeks.

Marketing and Storage

The three common methods for wrapping *ogiri* are (a) the use of dried banana leaves *Musa sapientum,* (b) the use of fresh or smoked leaves of the plant *Newbouldia laevis*, and (c) the use of small plastic wraps.

Leaf and plastic-wrapped *ogiri* samples bought from the local markets were examined immediately under a stereo microscope. Samples with no obvious fungal presence were selected. Three experimental designs were set up as follows: (a) a set of six samples (three from each type of leaf wrap) was smoked consistently for a week, (b) another set of six (two from each type of leaf and plastic wrap) remained unsmoked and stored at room temperature, and (c) the three types of wraps (minus *ogiri*) were placed in sterile plastic petri dishes and stored at room temperature.

Determination of Mycotoxins

Twenty gram samples from each experimental design (jute and nylon fiber bags) were analyzed for aflatoxin using the method of Kellert and

Spott (9). The modified method of Nowotny et al. (10) was used to screen 10-g samples for the other mycotoxins.

RESULTS

The use of clean dry nylon fiber bags proved very effective. Fermentation was observed to last 3 or 4 days. No fungal growth was noticed on the outside of the bag or on the fermented product even on day 3 before smoke treatment.

Using jute bags, fermentation lasted 5 to 6 days, and evidence of fungal contamination was obvious between days 2 and 3 of the fermentation. But when the jute bags received smoke treatment from day 2 to the final day of fermentation, no fungal contamination was observed. Whitish threads observed on jute bags on day 3 disappeared when smoke treatment was applied. The use of plastic bowls for fermentation was highly unsuitable because the process took longer—2 weeks.

When *ogiri* was smoked for 2 weeks, it had a very appealing aroma and texture. In contrast, the end product from the plastic bowl experiment lacked the characteristic *ogiri* aroma. When *ogiri* samples from both the jute and nylon fiber bags were assayed for mycotoxins, there was no evidence of contamination.

Effect of the Types of Wraps

Samples wrapped in dry leaves of the banana plant were less susceptible to fungal attack than *ogiri* wrapped in leaves of *Newbouldia laevis*. However, regular smoke treatment reduced the incidence of fungal contamination of *ogiri* in both types of leaf wraps. Plastic-wrapped samples had no observable fungi even up to 2 weeks of incubation but were devoid of the pleasant aroma characteristic of the smoked product.

DISCUSSION

It has been clearly demonstrated in this study that the use of clean dry nylon fiber bags instead of jute bags for the fermentation and early smoke treatment of the fermenting mash contributed significantly to the exclusion of fungi and thereby reduced the risk of mycotoxin contamination during *ogiri* production. Further related studies on methods of improving fermentation techniques on other products are now being considered.

REFERENCES

1. Jonsyn, F. E. 1988. Mycopathologia 104:123–127.

2. Jonsyn, F. E. 1989. Mircen Journal 5:547–562.

3. Jonsyn, F. E. 1990. Mycopathologia 110:113–117.

4. Jonsyn, F. E. 1991. In press.

5. H. G. Muller, personal communication.

6. Ogunsanwo, B. M., O. O. Faboya, O. R. Idowo, T. Ikotun, and D. A. Akano. 1989. Die Nahrung 33:983-988.

7. Odunfa, S. A. 1981. Journal of Plant Foods 3:245–250.

8. Antai, S. P., and M. H. Ibarahim. 1986. Journal of Applied Bacteriology 61:145–148.

9. Kellert, M., and H. J. Spott. 1980. Bundesgesundheitsblatt 23(1/2):13–21.

10. Nowotny, P., W. Baltes, W. Kroenert, and R. Weber. 1983. Chemie Mikrobiologie Technologie Der Lebersmitteln 8:24–28.

VII. COMMERCIALIZATION

24

Commercialization of Fermented Foods in Sub-Saharan Africa

Nduka Okafor

Fermented foods form an important part of the diets of people throughout the world, and the people of sub-Saharan Africa are no exception. In many parts of the world, as urbanization increases, the preparation of fermented foods moves from the small-scale household level to large-scale operations. Under these new conditions the foods are prepared with better scientific knowledge. For this reason large-scale factory procedures may differ from traditional approaches. For example, cheese that used to be produced with protease present in rennet may now be produced with protease produced by fungi.

With this in mind, a review was carried out in 1981 (1) to learn the extent to which some important fermented foods of sub-Saharan Africa had progressed toward commercialization. The stage that each food had attained was measured on a scale of 8, as shown in Table 1.

The purpose of this paper is to indicate to what extent various sub-Saharan fermented foods have progressed in the past decade toward being industrialized and to examine the role, if any, that modern techniques of biotechnology, particularly genetic engineering, have played in commercialization.

INDUSTRIALIZATION OF FERMENTED FOODS

Table 1 lists the fermented foods about which information is available, including those reviewed earlier (1). A review of the extent of progress toward industrialization of alcoholic beverages of sub-Saharan Africa was recently published (2) and is incorporated here into Table 1.

The following conclusions can be drawn:

- In 1981 the following foods had been produced on an industrial or

TABLE 1 Fermented Foods of Africa South of the Sahara

Food	Region	Processing	Level of Advance		Microorganisms
			1981 and	1991	
CASSAVA-BASED					
Garri	West Africa; Zaire	Pulp fermented	1,4,6,7	8	
Foo-foo (4)	Nigeria	Whole roots fermented	0	1	Cornebacterium Bacillus Lactic acid bacteria
Chikwangue	Zaire	Whole roots fermented	0		—
Lafun	Nigeria	Flour from chips	0		—
Kokonte	Ghana	Flour from chips	0		—
Cingwada	East, Central & South Africa	Flour from chips	0		—
CEREAL-BASED					
NON-ALCOHOLIC					
Ogi	Nigeria, Benin Republic	Fermented ground cereal	1,2,4,6	8 7,(8?)	
Koko (aflate) (5)	Ghana	Fermented ground cereal	1	1	Lactic acid bacteria
Mahewu (Mogow)	South Africa	Fermented ground cereal	1,2,4,5, 6,7,8		
Injera (10)	Ethiopia	Fermented ground cereal	1,2	1,2	Entero bacteria ceae, Lactic acid bacteria
MILK-BASED					
Ayib (16)	Ethiopia	Cheese-like	—	1,2	Lactic acid bacteria and yeasts
Nono	Nigeria	Fermented milk	—	1	*Lactobacillus bulgaricus* *L. plantarum,* *L. helveticus* *Streptococcus cremoris*
Fermented milk (3)	Zimbabwe	Fermented milk	—	1,2	*Lactococcus spp.*
"Lacto" (3)	Zimbabwe	Fermented milk	—	8	*Lactococcus spp.*
ALCOHOLIC					
Burukutu/Pito	West Africa	Fermentation of malted sorghum	1,2		
Sorghum (Kaffir) beer	South Africa	Fermentation of malted sorghum	1,2,4,5, 6, 7, 8		
Merissa (2)	Sudan	Fermentation of malted sorghum		0	
Bussa (2)	Kenya	Fermentation of malted sorghum		1,2	
PALM-BASED					
Palm wines	East, West, Central and South Africa	Spontaneous fermentation of palm sap	1,2,7		

MISCELLANEOUS					
Iru (dawadawa) (10)	Nigeria	Fermented seeds of Parkia	0	8	Lactic acid bacteria
Ogili (17)	Nigeria	Fermented seeds of castor oil	0	1,2	Lactic acid bacteria
Ugba (Ukraka) (6)	Nigeria	Fermented seeds of oil-bean	0	1,2	Bacillus
Fura (Ghussab)	Mali	Millet and cheese	0		—
Asami	East, Central South Africa	Fermented milk	0		—

Key: 1 = Organisms isolated
2 = Role(s) of organism(s) determined
3 = Selection and genetic improvement of organisms
5 = Improvement in raw material used
6 = Laboratory simulation of fermented food production
7 = Pilot plant production
8 = Industrial plant production

semiindustrial scale: *ogi*, *garri*, palm wine, *mahewu*, and sorghum (kaffir) beer.

Two new products are now being produced on an industrial or semiindustrial scale. The first is a Nigerian condiment known as *dawa-dawa*. It is being produced under the trade name of Dadwa by the firm of Cadburys in Nigeria from *Parkia* seeds as in the traditional fermentation. The second is a Zimbabwean fermented milk product known as Lacto. It is similar to the traditional fermented milk of Zimbabwe (3).

• The organisms involved in the fermentation of several foods that were unknown in 1981 have now been identified. They are *foo-foo* (4), *kokonte* (5), *ugba* (*ukpaka*) (6), and *ogili* (7,8).

The case of *dawa-dawa* is interesting. In 1981 the organisms involved were unknown; in 1991 not only are they known (9), but the food itself has been commercialized.

• Some foods not previously recorded have been added: *tej* from Ethiopia (10); *nono*, a milk-based product from Nigeria; and Zimbabwean fermented milk (3).

DISCUSSION AND CONCLUSIONS

As can be seen, very little has changed in the progress of the fermented foods of Africa toward industrial production. The 1990s are the era of biotechnology, especially genetic engineering. Fermented foods are brought about by microorganisms, and one would expect that these organisms would be subjected to the technology of gene cloning to improve their activity in the fermentation of foods.

For example, the fermentation of most carbohydrate foods such as cassava or maize is brought about by lactic acid bacteria. One would therefore have expected that these organisms would be targeted for improvement by gene cloning. Only one example of the advantage of the use of this technique will be given.

In *garri* fermentation lactic acid bacteria play an important part in producing the flavor of the food (11). Yet these organisms cannot split starch. If the amylase gene can be cloned into a lactic acid bacterium involved in *garri* fermentation, it is conceivable that fermentation will occur faster. If the gene for linamarase production can also be simultaneously cloned, then not only will the fermentation be faster but detoxification also will occur (12).

The only work having any relationship to gene cloning in organisms involved in fermentation was the isolation of plasmids from cassava fermenting organisms by Nwankwo et al. (13). They found that they could not transfer the plasmids to *E. coli* and there the work ended.

The lack of ability to exploit this new technique in an area of vital importance to Africa south of the Sahara is a clear example of (an almost?) missed opportunity in an age when seemingly everyone is cloning a gene from one source or another. Nevertheless, there have been some developments in other directions. For example, Ofuya and Nnajiofor (14) have developed a starter culture for *garri* that should prove useful in the commercialization of the food. Also, Ofuya and Fiito (15) have developed a rapid method for assessing the quality of *garri* based on an iodine reaction.

REFERENCES

1. Okafor, N. 1981. A scheme for the improvement of fermented foods of Africa, south of the Sahara. Pp. 61–69. In: Global Impacts of Applied Microbiology. S. O. Emejuaiwe, O. Ogunbi, and S. O. Sanni (Eds.). London: Academic Press.

2. Okafor, N. 1990. Traditional alcoholic beverages of tropical Africa: Strategies for scale-up. Process Biochemistry International 25:213–220.

3. Feresu, S. B., and M. I. Muzondo. 1990. Identification of some lactic acid bacteria from two Zimbabwean fermented milk products. World Journal of Microbial Biotechnology 6:178–186.

4. Okafor, N., C. O. Oyolu, and B. C. Ijioma. 1984. Microbiology and biochemistry of *foo-foo* production. Journal of Applied Microbiology 55:1–13.

5. Mensah, P., A. M. Tomkins, B.S. Drasar, and T. J. Harrison.

1991. Antimicrobial effects of fermented Ghanaian maize dough. Journal of Applied Bacteriology 70:203–210.

6. Obeta, J. A. N. 1983. A note on the microorganisms associated with the fermentation of the seeds of the African oil bean tree. Journal of Applied Bacteriology 54:433–435.

7. Ogundana, S. K. 1980. The production of *ogiri*: Nigerian soup condiment. Lebensmittel Wissenschaff und Technologia 13:334–336.

8. Onunkwo, A. U. 1982. Some edible fermentation products of Nigeria. M.Sc. thesis, University of Strathclyde, Glasgow.

9. Odunfa, S. A. 1981. Microorganisms associated with the fermentation of the African locust bean, *Parkia filicoidea*, during *iru* preparation. Journal of Plant Foods 3:245–250.

10. Girma, M., B. A. Gashe, and B. Lakew. 1989. The effect of fermentation on the growth and survival of *Salmonella typhimurium, Staphylococcus aureus, Bacillus cereus*, and *Pseudomonas aeroginosa* in fermenting *tef (Eragrostis tef)*. Mircen Journal of Applied Microbiology 5:61–66.

11. Okafor, N., and J. Uzuegbu. 1987. Studies on the contributions of microorganisms on the organoleptic properties of *garri*, a fermented food derived from cassava *(Manihot esculenta* Crantz). Journal of Food Agriculture 2:99–105.

12. Okafor, N., and A. O. Ejiofor. 1990. Rapid detoxification of cassava mash fermenting for *garri* production following inoculation by a yeast simultaneously producing linamarase and amylase. Process Biochemistry International 25:82–86.

13. Nwankwo, D., E. Anadu, and R. Usoro. 1989. Cassava fermenting organisms. Mircen Journal of Applied Microbiology 5:169–179.

14. Ofuya, C. O., and C. Nnajiofor. 1989. Development and evaluation of a starter culture for the industrial production of *garri*. Journal of Applied Microbiology 66:37–42.

15. Ofuya, C. O., and J. Fiito. 1989. A rapid method for determining the quality of *garri* based on iodine reduction test. Letters in Applied Microbiology 9:153–155.

16. Ashenafi, M. 1990. Effect of curd cooking temperatures on the microbiological qualities of *ayib*, a traditional cottage cheese. World Journal of Microbial Biotechnology 6:159–162.

17. Odibo, F. J. C., and A. I. Umeh. 1989. Microbiology of the fermentation of Telfaria seeds for *ogiri* production. Mircen Journal of Applied Microbiology and Biotechnology 5:217–222.

25

Biotechnology for Production of Fruits, Wines, and Alcohol

J. Maud Kordylas

Fermentation is biotechnology in which desirable microorganisms are used in the production of value-added products of commercial importance. Fermentation occurs in nature in any sugar-containing mash from fruit, berries, honey, or sap tapped from palms. If left exposed in a warm atmosphere, airborne yeasts act on the sugar to convert it into alcohol and carbon dioxide. The making of wines and beers uses this biotechnology under controlled conditions. Alcoholic beverages have been produced for centuries in various societies. They are often central to the most valued personal and social ceremonies of both modern and less literate societies. In such traditional ceremonies as childnaming, marriage feasts, and funerals, alcoholic beverages are often present. In Africa, maize, millet, bananas, honey, palm and bamboo saps, and many fruits are used to ferment nutrient beers and wines. The best known being kaffir beer and palm wines.

Industrial fermentation processes are conducted with selected microorganisms under specified conditions with carefully adjusted nutrient concentrations. The products of fermentation are many: alcohol, glycerol, and carbon dioxide are obtained from yeast fermentation of various sugars. Butyl alcohol, acetone, lactic acid, monosodium glutamate, and acetic acid are products of bacteria action; citric acid, gluconic acid, antibiotics, vitamin B_{12}, and riboflavin are some of the products obtained from mold fermentation.

YEASTS

Yeasts, the main microorganisms involved in alcoholic fermentation, are found throughout the world. More than 8,000 strains of this vegetative microorganism have been classified. About 9 to 10 pure strains, with their subclassifications, are used for the fermentation of grain mashes. These belong to the type *Saccharomyces cerevisiae*. Each strain has its own characteristics and imparts its special properties to a distillate when used in fermentation. A limited number of yeasts in the classification *Saccharomyces ellipsoides* are used in the fermentation of wines from which brandy is distilled. The strains used in the fermentation of grain mashes are also used in the fermentation of rum from sugarcane extracts and in beer production. Since yeasts function best in slightly acid medium, the mash, juice, sap, or extract prepared for fermentation must be checked for adequate acidity. If acidity is insufficient, acid or acid-bearing material are added. For distilled liquors, fermentation is carried out at 24° to 29°C for 48 to 96 hours, when the mash or must is ready for distillation. The alcohol content of the fermented must is about 7 to 9 percent.

RAW MATERIALS

Cereals and Starchy Roots

For most distilled liquors, the raw material used is a natural sugar as found in honey, ripe fruit, sugarcane juice, palm sap, beet root, milk, or a substance of amylaceous (starchy) nature that can be easily converted into simple sugars using enzymes present in cereals or through the addition of suitable malted cereal. Maize or corn is the most important grain used as fermentable starchy cereal. Starchy roots and tubers are also used. Industrial production of alcohol from cassava in Brazil has been described by De Menezee (1). The alcohol produced is concentrated in a second distillation column to 97.2 percent and is further dried to 99.9 percent and blended with gasoline for energy purposes.

Malt is important in distilled liquor. In addition to converting starches from other carbohydrates to sugars, malt contains soluble proteins that contribute flavor to the distillate obtained from the fermentation of grain malt mixtures.

Sugarcane

Grown throughout the tropics and semitropics, sugarcane and its products, including cane juices, molasses, and sugar are used to make rum and an alcohol derived from rum. Pressed juice from sugarcane can be used as the base raw material for fermentation, or the juice can be concentrated for sugar production with the molasses residue from sugar crystallization used as a base for alcohol fermentation. Molasses contains about 35 percent sucrose and 15 percent reducing sugars. This gives molasses its principal value as an industrial raw material for fermentation to produce rum. Two or 3 liters of molasses produces 1 liter of rum. Acetone and butanol also are produced from molasses by fermentation with *Clostridium* bacteria. Food yeast *Torulopsis utilis*, is prepared from molasses, as are baker's and brewer's yeasts (2).

Coconut Palm

The coconut palm finds many uses on the tropical islands of the Pacific. Toddy is produced by tapping the unopened flower spathe of the coconut palm. The spathe is bruised slightly by gentle tapping with a small mallet and is tied tightly with fiber to prevent it from opening. It is bent over gradually to allow the toddy to flow into a receptacle. About 5 centimeters is cut from the tip of the spathe after about 3 weeks. Thereafter, a thin slice is shaved off once or twice a day and the exuding sap is collected. Palms are tapped for 8 months of the year and rested for 4 months. The average daily yield per palm is about 2 liters. The yield per spathe varies from 15 to 80 liters, and an average palm can yield 270 liters during 8 months of tapping. The fresh sweet toddy contains 15 to 20 percent total solids, of which 12 to 17.5 percent is sucrose.

Toddy ferments rapidly due to naturally occurring yeasts. Fermented toddy contains about 6 percent alcohol. After 24 hours the toddy contains 4 to 5 percent acetic acid and is unpalatable as a beverage. It can be used for the production of vinegar. Fermented toddy can be distilled to produce arrack. Freshly fermented toddy is used instead of yeast in bread making. Constant tapping of coconut palms for toddy eliminates the nut crop. In 1952 in wine distilleries in Sri Lanka, over 49 million liters of toddy was fermented to give 4.5 million proof liters of arrack (2).

Oil Palm

By tapping the male inflorescence of the oil palm, a sweet sap is obtained. The leaf subtending the immature male inflorescence is removed to provide access, the inflorescence is excised, and thin slices are cut once or twice daily. The exuding sap is funneled into a calabash or a bottle. The fresh sap contains 15 percent sugar. Tapping is done daily for 2 to 3 months, yielding about 3.5 liters of sap per day. The sap ferments by the action of bacteria and natural yeast to produce a beverage with a milky flocculent appearance and a slight sulfurous odor known as palm wine. Palm wine is produced and marketed in considerable quantities in Nigeria.

The sap may be boiled to produce dark-colored sticky sugar or jaggery, which does not keep well. About 9 liters of juice produces 1 kilogram of jaggery. The fermented sap also yields yeasts and vinegar. A mean annual yield of 4,000 liters of sap per hectare of 150 palms has been recorded in eastern Nigeria. This was estimated to have a value more than double that of oil and kernels from similar palms. Tapping, however, reduces the fruit yield. Sap can also be obtained by tapping the crown of the tree laterally or by felling the palm and drilling a hole through the growing point. Both these methods are very wasteful since they kill the plant. The Palmyra palm yields about 2 liters of palm sap per day. Large palms with several tapped inflorescences give as much as 20 liters per day. A single palm of this type is estimated to produce 12,000 liters of sap during its tapping life.

Fruits

Grapes are the most common fruit used as raw material for alcoholic fermentation. They are used in distilled liquor to make brandy. Historically, wine is the product of fermentation of grape species *Vitis vinifera*. The high sugar content of most *V. vinifera* varieties at maturity is the major factor in their selection for use in much of the world's wine production. Their natural sugar content provides the necessary material for fermentation. It is sufficient to produce a wine with an alcohol content of 10 percent or higher. Wines containing less alcohol are unstable because of their sensitivity to bacterial spoilage. The grape's moderate acidity when ripe is also favorable to wine making. The fruit has an acidity of less than 1 percent, calculated as tartaric acid, the main acid in grapes, with a pH of 3.1 to 3.7. The flavor of grapes varies from neutral to strongly aromatic, and the pigment pattern of the skin varies from light greenish-yellow to russet, pink, red, reddish violet, or blue-black. Grapes also contain tannins needed to give bite and taste in the flavor of wines and to protect them from bacteria and possible ill effects if overexposed to the air.

Other fruits can be used to produce wine. When fruits other than grapes are used, the name of the fruit is included, as in papaya or pineapple wine. Apples and citrus fruits with sufficient fermentable sugars are crushed, and the fermentable juices are either pressed out for fermentation or the entire mass is fermented. Tropical fruits such as guava, mangos, pineapple, pawpaw, ripe banana, ripe plantain, tangerine, and cashew fruit also contain fermentable sugars with levels varying from 10 to 20 percent. Overripe plantain pulp was reported to contain 16 to 17 percent fermentable sugar, with the skin containing as much as 30 percent (3).

The tropical climate prevailing in Africa is ideal for the growth and multiplication of microorganisms. The environment is abundant in biomass and in raw materials, which are high in starches and sugars and can be used for fermentation. The available literature is sufficient in information on conditions and control measures required for optimum microbial activity in the various microbial processes. Convincing research results are also available to support utilization of microorganisms in the production of high-quality products of commercial importance. What is lacking, however, is organization of the available information to enable selection of appropriate microbial processes that can be put together to form an integrated system to harness desirable microorganisms as a labor force for industrial exploitation. Below an account is given of an attempt to organize four microbial processes into a production system to produce fruits, wines, and alcohol in an experimental project.

INTEGRATED PRODUCTION SYSTEM

An experimental project was established aimed at providing adequate conditions and control measures in four separate biological subsettings to produce quality products through the action of microorganisms. An attempt was then made to synchronize the activities of the subsettings into an integrated system for the production of fruits, wines, and alcohol with jam production as an integral part of the production system.

The four biotechnological subsettings used were: a compost pile, stimulated microbiological activity in the soil for release of nutrients, yeast activity in extracted fruit juices for the production of wines, and yeast activity in juice extracted from pineapple by-products for the production of alcohol.

Composting

In 1984 a two-compartment wooden structure measuring 2 × 1 × 1 meters was constructed to hold two piles of composting material. Cut grass, straw, dried leaves, and other high-carbon organic wastes were collected from the neighborhood. They were layered with chicken manure to provide a nitrogen source to form compost piles within the compartments. Kitchen waste and, later, wastes from fruit processing were also added to the piles. The piles were kept sufficiently moist by sprinkling with water. To encourage optimum microbiological activity, the piles were aerated by constant turning. Observation of heat generation and the rates at which the piles were digested were used to indicate effective microbial activity. The lack of offensive odor from the piles was considered a sign of adequate control conditions within the piles.

Microbial Activity in Soil

The compost obtained was used to prepare selected sites in a backyard plot measuring 9 × 20 meters that was originally filled with clay soil. The clay soil was removed, and mixed with compost. The mixture was placed into the holes to form raised beds for planting. Two guava seedlings obtained from the research station at Njombe were added to other fruit seedlings nursed in pots. These were transplanted into the prepared sites. As more compost was made available, more fruit seedlings were transplanted into position. By mid-1986 the backyard plot was planted with the following fruit trees: six soursops, five guavas, three pawpaw, eight carambola bushes, one mango, and one avocado pear. The fruit trees were interplanted with plantains, cocoyam, pepper, and a few winged bean plants to form a multistory system as usually obtained in traditional cropping systems in Africa.

Sufficient compost was applied regularly to the soil to encourage microorganisms and other soil dwellers to function and to enhance mycorrhizal fungi association with root hairs, to provide nourishment and protection and for the well-being of the plants. The compost was applied by removing the topsoil around the plant to expose the roots. Two to three loads of compost were distributed evenly around the roots and were covered with the topsoil. Fallen leaves around the yard were raked and used as mulch to cover the top of the disturbed soil to prevent it from eroding away during heavy rains. The leaf mulch was also used to protect the soil surface from the pounding rains. It also kept the soil cool during the dry season and helped to conserve soil moisture when the plants are irrigated. To encourage microbial activity in the soil, no inorganic fertilizer was applied and no pesticides were sprayed anywhere in the yard.

The fertility of the soil around the growing plants was regularly monitored using a two-prong fertilizer analyzer that indicated whether the soil had sufficient nitrogen, potassium, and phosphorus. Where a deficiency was indicated, more compost was applied to the soil. The method of removing the topsoil to apply compost aerated the soil. During the rainy season the edges of the soil around the raised beds were lifted slightly with a fork to allow air in without disturbing the soil. The improvement in soil fertility over the years, the physical appearance of the growing trees, the lack of disease, and later the fruit yield were used as parameters to indicate optimum conditions in the soil that promoted microbial activity. Fruit harvests were recorded daily.

Wine from Fruit Juices

Extracted juices from pawpaw and carambola harvested from the backyard and juice extracted from pineapples obtained from the local market were used to carry out wine-making experiments. The pulp remaining after juice extraction from fruits was used to make jam.

To prevent the growth of undesirable microorganisms, the juice extracts were pasteurized. All utensils, tools, and equipment that came into contact with the wine in making, were sterilized and rinsed thoroughly. No chemicals were used in the preparation of the must. Sufficient amounts of yeast nutrients were added for yeast growth. The pH of the must was adjusted and sufficient sugar was added where needed to produce 11 percent alcohol in the finished wine. A small amount of tannin solution was added to provide bite and flavor to the finished wine. The yeasts used for the first experiments were activated according to the manufacturer's directions. Thereafter, pawpaw, pineapple, and carambola wine yeasts were reserved from wines made. These were kept under refrigeration and used for subsequent wine production. All the wine-making stages—first and second fermentations, raking, storage and aging—were carried out in an air-conditioned room so that constant temperatures could be maintained. Finished wines were bottled, pasteurized, cooled, and corked for storage to age in the bottles.

Alcohol Production from Pineapple

The preparation of pineapples usually produced about 40 to 50 percent waste materials. This was made up of the top crown, the fibrous outside skin, the seeded inner cover, and the hard central core. The crown and the fibrous skin were added to the compost pile. The seeded cover and the central core were crushed and kept frozen until needed for juice extraction for fermentation. The sugar level of the pasteurized juice was checked and sufficient amounts of granulated sugar were added to produce about 12 percent alcohol in the fermented must. The pH of the preparation was also adjusted. The fermented must was then distilled. The temperature of the distillation was carefully controlled so that a high concentration of alcohol could be obtained from one distillation. The bulk of the alcohol collected was over 90 percent concentration. This alcohol was used in experiments with fruits to make aperitif drinks and liquors.

INTEGRATION

The activities of the four microbial processes were synchronized and integrated into an interdependent production system where the subprocesses provided support for each other. The composting setup received wastes from fruit processing. The compost was used to enrich the soil in which the fruit trees were planted. Harvested fruits provided juice extracts for wine making, and by-products from fruit processing provided raw materials for alcohol production. Jams were produced from fruit pulp and were marketed to provide financial support for needed research and to purchase equipment.

RESULTS AND DISCUSSION

Composting

It took about 12 months of composting to arrive at the number of turnings needed, and the correct ratios of high-carbon materials to nitrogenous material required to prepare a compost pile without an ammonia odor. When the correct proportions were used, the compost was completed within 3 weeks during the hot dry weather, and in 4 to 5 weeks during the cool rainy season. Sufficient heat was generated to sterilize the compost, and no odor was detected.

Soil and Fruit Production

It took 2 to 3 years of regular application of compost for the clay in the planted sites to change into dark fluffy soil. Earthworms were seen in the soil after 3 to 4 applications of compost. During the first 3 years the growing plants were constantly affected by plant diseases. The infections diminished, however, as the soil fertility improved. None of the infections were serious enough to require action. The attacks increased during the dry season and again toward the end of the rains, especially during periods when the rains were long and heavy.

Table 1 shows guava, soursop, and carambola yields over the years. After their first bearings, most of the trees lost their seasonality and continued to flower, set fruit, mature, and ripen fruit as long as the weather and soil conditions remained favorable. The rains usually started in March/April and enhanced fruit yield. Thereafter, fruit yields were affected by how heavy the rainy season was and how long it lasted. Flowering and fruit settings were greatly diminished in the guava and the soursop during heavy rains. They were, however, resumed as soon as there was a break in the rains. The next harvests were delayed if the rains were heavy and lasted for a long time. The

TABLE 1 Fruit Yields (kilograms), 1986–1991

	Guava				Soursop				Carambola			
Year	Jan.-June	July-Dec.	Total	Ave./Tree	Jan.-June	July-Dec.	Total	Ave./Tree	Jan.-June	July-Dec.	Total	Ave./Tree
1986	7.2(2)	41.9(2)	49.0(2)	24.50	—	—	—	—	—	—	—	—
1987	54.8(2)	76.4(2)	131.2(2)	65.6	163.2(4)	9.9(2)	173.0(4)	43.3	—	0.4(1)	0.4(1)	4.0
1988	86.4(3)	131.1(3)	217.5(3)	72.5	132.4(4)	19.2(4)	151.5(4)	37.9	17.0(6)	46.6(6)	63.6(6)	10.6
1989	109.9(3)	98.5(4)	208.3(4)	52.1	294.1(6)	105.6(5)	394.0(6)	66.2	86.2(8)	135.3(8)	221.5(8)	27.7
1990	165.5(5)	129.5(5)	295.0(5)	59.0	195.7(6)	92.6(5)	286.3(6)	47.7	143.5(8)	135.7(8)	279.0(8)	34.9
1991		116.6(5)			341.2(6)				154.9(8)			

( ), number of trees bearing fruit.

carambola somehow continued to flower and set fruit during the rainy season as long as there was periodic sunlight.

Quality was high in guavas and soursop harvested at the beginning of the rains. The fruits were large and well formed and had good flavor. Most of the fruits harvested at the ends of the dry and rainy seasons were smaller, malformed, or diseased. This may be due to the effects of too little or too much water on the health of the plants. Too little water may have affected the activities of microorganisms in the soil, and too much water may have reduced air supply to microorganisms in the soil and leaching of nutrients from the soil. Diminished microbial activity may have affected the well-being of the plants. These assumptions might, however, need to be confirmed through controlled experiments.

The 180-square meter backyard plot yielded sufficient quantities of fruits—guava, soursop, and carambola—to provide raw materials for processing to make jams available on the local market throughout 1989 and thereafter. Carambola yields were also sufficient for wine making. The amount of pawpaw harvested from the backyard was not sufficient, however, for both jam production and wine making. More pawpaw was therefore purchased from the local market to supplement the amount harvested. The quantity of mango obtained from the one mango tree was also not sufficient to keep up with the demand for mango jam on the market. More was obtained from the local market.

Table 2 shows total yields for guavȧ, soursop, carambola, and pawpaw harvested from 1986 to 1990. Although two of the four pawpaw trees died, total yields of fruits from the backyard continued to increase over the years. Yields from crops interplanted among the fruit trees, including pepper, cocoyam, plantain, and winged beans, and from the one avocado tree that started bearing fruit in 1990, when added to those obtained from trees in Table 2, provided an overall yield of over 1 ton from the backyard plot in 1989 and again in 1990.

Wine Production

Wine of acceptable quality were produced from pawpaw, pineapple, and carambola. The wines made were either dry, semidry, or sweet.

TABLE 2 Fruit Yields (kilograms), 1986–1990

Fruit	1986	1987	1988	1989	1990
Guava	49.0	131.2	217.5	208.3	295.0
Soursop	—	173.0	151.5	397.0	286.3
Carambola	—	0.4	63.6	221.5	279.0
Pawpaw	—	28.3	100.9	72.6	40.0
Total	49.0	332.9	533.5	899.4	900.3

Although no controlled organoleptic assessment was organized to evaluate the acceptability of the wines, reactions from random individuals who tasted the wines were favorable. Marketing trials will be conducted.

Alcohol Production

Juice extracted from the crushed pineapple core and the inner seeded cover contained sufficient sugar to produce 6.5 to 7 percent alcohol after fermentation. With the addition of extra sugar, however, the alcohol content was increased to 10 percent. A total of 25 liters of over 90 percent concentration alcohol was distilled from 200 liters of discarded wines and 100 liters of fermented pineapple waste extract. Portions of the alcohol were used to carry out experiments to produce aperitif drinks with guava, pineapple, passion fruit, carambola, and ginger. The experiments are still in progress.

BIOTECHNOLOGY PRODUCTION SYSTEM

The integrated bioechnology research and development system is shown in Figure 1. The broken-line arrows indicate units not yet included but for which information has been collected to enable their future integration into the system. The chickens are needed to produce manure for the composting process, with meat and eggs as additional marketable products. Wastewater from fruit processing would be recycled to provide water for irrigation and for composting to economize on the use of potable water for those processes.

From the data collected and from experience gained through the project, the integrated biotechnology production system has many advantages:

- It is environmentally sound: Wastes generated from fruit processing and from the backyard plot are recycled through the composting process to produce organic fertilizer.
- Labor requirements have not been excessive: Once the necessary conditions are met and controls applied for microorganisms to grow and multiply, the productive processes for wine and alcohol production, for composting, and for nutrient release for plant nourishment are carried out with little or no supervision.
- Energy requirements are low: Apart from the energy needed for production of jams and for pasteurization and to run the small-scale equipment used in processing, the integrated production system needs limited amounts of energy input to function. The microbial processes generate their own energy. The need for air conditioning to maintain

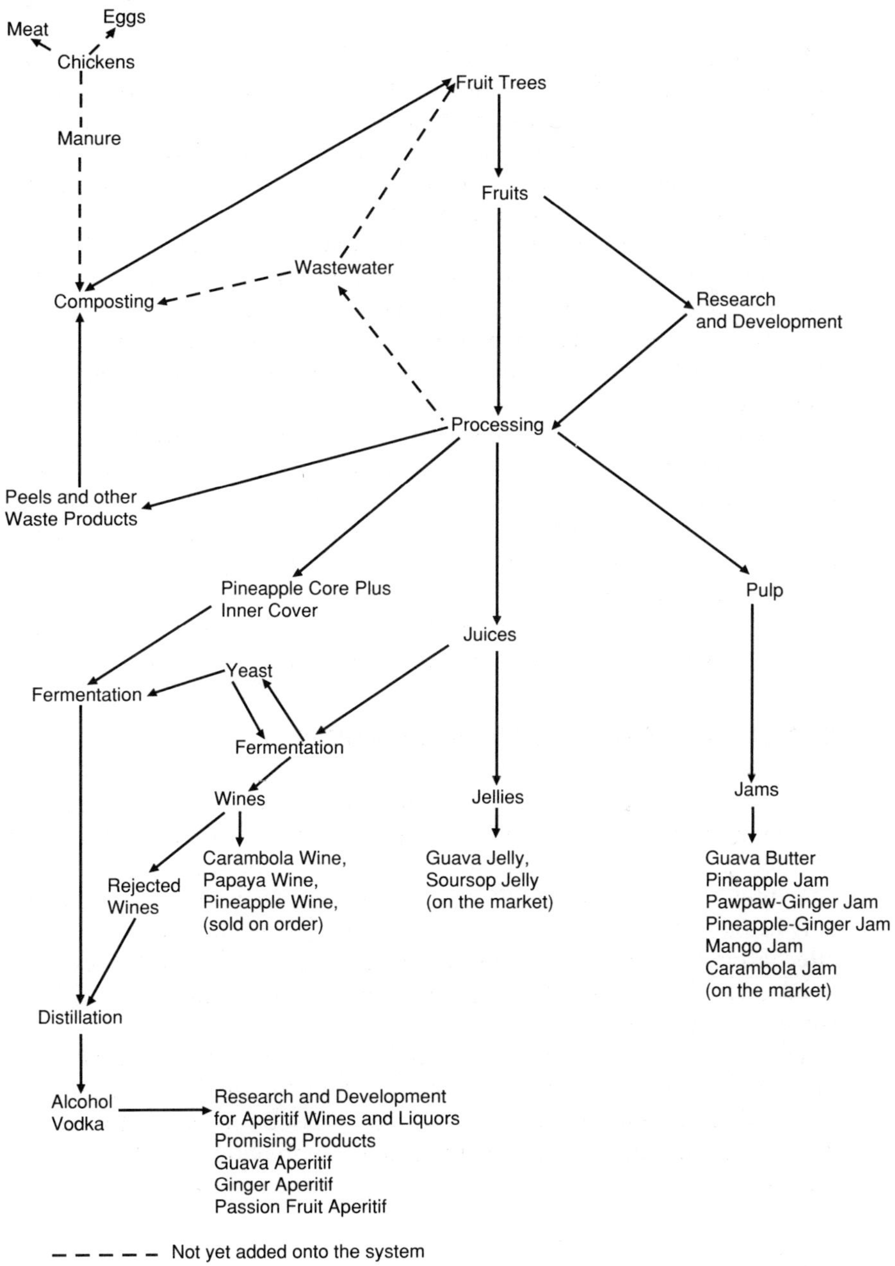

FIGURE 1 An integrated research and development system using biotechnology in the production of fruits, wines, and alcohol.

constant environmental temperatures will likely add to the energy costs.

- The system is sustainable: The interdependency of the microbial subprocesses provides sustainable support to each other with limited input required from outside. Funds generated from the sale of products (jams, wines, apertif drinks) are used to support needed research and to purchase equipment and supplementary produce required to sustain the production of marketable products.
- Only practical research is undertaken: Experiments carried out are those needed to solve immediate problems arising from the production system. These are carried out either to improve the quality of a product, to formulate new products from raw materials or by-products generated within the system, or to enhance marketability of a product.
- Realistic data is collected for feasibility reports. Production and trial marketing of products from the system have enabled real data to be collected. These are being used to evaluate the system economically and to produce a feasibility report based on actual figures to make decisions on establishing an industry based on the prototype research and development unit.
- Valuable experience has been gained: The project has provided valuable experience in the management of a small enterprise.

CONCLUSIONS AND RECOMMENDATIONS

A good number of efficient microbial processes are available. Sufficient knowledge has been accumulated and information provided on their management and control. If properly selected, synchronized, and integrated, the activities of microorganisms from such processes may be harnessed and used. Their exploitation may be a more promising alternative to large-scale industrial technologies imported from developed countries, which developing countries in Africa cannot afford, sustain, or manage.

The priority for research is, therefore, on selecting the right types of microbial processes that can be put together to form sustainable productive systems, with research trials carried out on prototypes to determine the most economically viable combinations to be adopted for commercial exploitation.

REFERENCES

1. De Menezee, I. J. B. 1978. Alcohol Production from Cassava. Pp. 41–45 in: Cassava Harvesting and Processing. International Development Research Center, Ottawa, Canada.

2. Purseglove, J. W. 1985. Tropical Crops. In: Monocotyledons. England: Longman.

3. Kuboye, A. O., A. B. Oniwinde, and I. A. Akinrele. 1978. Production of Alcoholic Beverages from Ripe Pineapples, Plantain, and Bananas, Vol. 2, Pp. 78–80. Nigerian Institute of Food Science and Technology. Lagos, Nigeria.

26
Future Directions

Leslie Fook-Min Yong

The preparation of fermented foods predates the recorded history of Man. Early humans used observation of the apparent effects of microbial alteration of food characteristics to develop processes for food fermentation. The resultant fermented products normally have a different texture and flavor compared to the unfermented starting materials, thus making them more palatable and digestible and prolonging their shelf life. Technical progress was initially slow, as reflected in the long fermentation periods required; it was incremental to the technical know-how and basic scientific information then available. It is probably fair to say that in the very early days brew-masters were more artisans than technologists. With the rapid advancement in understanding of the basic sciences of microbiology and biochemistry, coupled with the introduction of new equipment, the developed nations have forged ahead in improving the safety and efficiency of the bioprocesses used to manufacture traditional fermented foods, such as cheese fermentation.

"OLD" AND "NEW" BIOTECHNOLOGY

With the rapid progress in the biological sciences, both basic and applied aspects, it has been possible to gain a better understanding of the mystery that has surrounded fermentation processes. The types of microorganisms involved has been isolated and identified, and the physiology and metabolism of these organisms have been studied. Hence, traditional fermented foods can now be made better, faster, and more economically. The application of available knowledge to

---

In this paper I draw on my experience working with soy sauce fermentation and then proceed to discuss the production of flavor and fragrance materials by microbial fermentation. Experience gained from this traditional fermented condiment has enabled me to develop novel bioprocesses for the production of aroma chemicals.

improve traditional food fermentations in developed countries has far outpaced that in developing countries.

The terms "old biotechnology" and "new biotechnology" have been used—"old" to mean the undirected manipulation of microorganisms and plants, such as by mutagenesis and selection of the better strains. In this old biotechnology I would like, for convenience, to include directed control of the physical and chemical environments of the fermentation process, which could result in better performance of the useful microbes.

Though mutation increases the ability to select better strains, there can, of course, be little directed alteration of genetic material. The new biotechnology, such as recombinant DNA techniques, overcomes this problem. The new biotechnology can, of course, be of tremendous help in producing superstrains of microbes that could enable acceleration of fermentation processes, provide more efficient utilization of raw materials, and produce better-quality products. How best can developing nations apply these biotechnologies to traditional fermented foods? Should it be application of the "old" before the "new," "new" without the "old," or "old" and "new" simultaneously?

In their enthusiasm to promote the new biotechnology for traditional fermented food applications, scientists from developed countries should not forget the different environments that exist in developed and developing countries. In developed countries the old biotechnology is already well understood and practiced efficiently in fermented food industries. Developing countries may need to acquire a better understanding of the old biotechnology before efficiently absorbing and implementing the new biotechnology to its fullest.

APPLICATION OF BIOTECHNOLOGY

Preparation of traditional fermented foods is more complex and time consuming than that involved in the production of single chemical substances. For example, in soy sauce fermentation more than one type of microorganism is involved, whereas in citric acid fermentation only one species of fungus is normally used. How can developing countries apply new knowledge in the old and new biotechnologies to their own complex traditional food fermentations?

Take soy sauce fermentation as an example of a traditional fermentation process conducted in a developed country, such as Japan compared with that in a country like Malaysia. The technology in use in Japan is sophisticated, very advanced, and highly productive and mechanized. The microbes used have been selected over the years for their performance in producing a better-quality product. The cottage industry soy sauce fermentation in Malaysia is highly labor intensive and usually relies on "natural" inoculation of raw materials using unwashed trays for previous fermentations rather than using a separately prepared inoculum of *Aspergillus oryzae*.

The equipment used in Japan to conduct the fermentation is state-of-the-art machinery with microprocessor or computer control to provide the optimum conditions for microbial growth and activity. The microorganisms used have been manipulated by mutagenesis to give better performance, such as better enzymatic activity to give better hydrolysis of proteinaceous matter in defatted soybean meal as well as better flavor production. In comparison, the average process used in Malaysia could be considered primitive.

This disparity is attributable to a better understanding of the theoretical and practical bases of soy sauce fermentation by scientists and technicians in Japan's soy sauce factories. The old biotechnology involved in this type of traditional fermentation is well understood in Japan, and the Japanese are now able to make better use of the new biotechnology—such as the directed alteration of genetic material of the mold (*Aspergillus oryzae*), yeast (*Saccharomyces rouxii*), and bacteria (*Pediococci*) used in soy sauce fermentation so as to improve their fermentative qualities.

Necessary Prerequisites

For developing countries to make full use of the available biotechnologies in their traditional food fermentations, an understanding and acquisition of expertise in the following areas are essential.

Art of fermentation

A clear understanding by the master brewer of every step used in the fermentation is needed. This is the art of fermentation. Although the master brewers might not have scientific backgrounds, they could normally ensure a proper fermentation as a result of years of experience. Without a knowledge of the art of traditional food fermentation, a scientist cannot provide a scientific explanation for the process and proceed to provide assistance in improvement of the process.

Microbiology

It is essential to know which microorganisms involved in food fermentations are useful and how the physiology and metabolism of these microbes are affected by the physical and chemical environments of fermentations, as well as how their microbial activities in turn affect the fermentation processes. Microorganisms normally break down carbohydrates, proteins, and lipids present in the raw materials to be fermented by releasing enzymes into the medium. As the raw materials are hydrolyzed, the environment is changed, as sometimes reflected by a drop in pH value. Moreover, the breakdown products such as peptides and amino acids can be further converted into smaller volatile molecules that are odoriferous and hence improve the flavor characteristics of the fermented foods.

Upstream and downstream processing

Normally raw materials are pretreated before fermentation. It is important to comprehend how such pretreatment could affect the fermentation process. In soy sauce fermentation, whole soybeans are steamed to make the soy protein more easily hydrolyzable by the proteases of *Aspergillus oryzae*. In so doing, too much moisture is introduced and wheat flour must be added to lower the moisture content to a level that does not favor early bacterial growth and hence prevents spoilage of the fermentation.

Downstream processing does not affect the bioprocess involved. However, it could alter the normal organoleptic properties of the product, especially when downstream processing involves heating, such as in the pasteurization of soy sauce. Heating causes a change in the flavor of soy sauce due to nonenzymic browning reactions, which could result in the production of pyrazine compounds.

Biochemistry

An understanding of the biochemical activities of the microbes actively participating in the fermentation could help to explain the change in the texture of the raw material as well as the origin of flavoring substances often present in fermented foods. Flavor and texture are important properties of fermented foods. Elucidation of flavor production in such fermentations could result in the development of processes for producing of flavoring materials by fermentation, as in the production of cheese flavors by *Penicillium roquefortii*.

Fermentation equipment and techniques

Practical experience in the use of both solid-state and submerged culture fermentation equipment is very useful. Normal training includes submerged culture bioreactors but not solid-state fermenters. It is useful to know both types of fermentations because traditional food fermentations often involve solid state fermentation. In soy sauce fermentation an initial solid-state fermentation is followed by a submerged fermentation step. Systems that measure and control pH, dissolved oxygen, temperature, and moisture help to make these bioprocesses more efficient and reduce the time required for production of a quality product.

CONCLUSIONS

For developing countries, future directions in applying biotechnology to traditional fermented foods should be: (1) training of a pool of technicians in the art and science of traditional food fermentations and (2) investigations by local scientists into the scientific basis of indigenous food fermentations.

Theoretical basic science education, such as the microbiology and biochemistry of food fermentations, could be taught in schools; so could the use of modern bioreactor systems. However, the application of such biotechnological knowledge to actual commercial fermentations can come about only after a practical experience in a fermented food factory for a period of time. The approach to be taken in applying biotechnology to traditional food fermentations should be that of finding solutions to existing bioprocessing problems and not trying to find problems with newly acquired biotechniques.

Only after the old biotechniques of fermentation have been successfully used can industries in developing countries look forward to using the new biotechniques of recombinant DNA to improve the genetic constitution of the microorganisms involved.

Board on Science and Technology for International Development

Board on Science and Technology for International Development
Publications and Information Services (FO-2060Z)
Office of International Affairs
National Research Council
2101 Constitution Avenue, N.W.
Washington, D.C. 20418 USA

How to Order BOSTID Reports

BOSTID manages programs with developing countries on behalf of the U.S. National Research Council. Reports published by BOSTID are sponsored in most instances by the U.S. Agency for International Development. They are intended for distribution to readers in developing countries who are affiliated with governmental, educational, or research institutions, and who have professional interest in the subject areas treated by the reports.

BOSTID books are available from selected international distributors. For more efficient and expedient service, please place your order with your local distributor. Requestors from areas not yet represented by a distributor should send their orders directly to BOSTID at the above address.

Energy

33. **Alcohol Fuels: Options for Developing Countries.** 1983, 128 pp. Examines the potential for the production and utilization of alcohol fuels in developing countries. Includes information on various tropical crops and their conversion to alcohols through both traditional and novel processes. ISBN 0-309-04160-0.

36. **Producer Gas: Another Fuel for Motor Transport.** 1983, 112 pp. During World War II Europe and Asia used wood, charcoal, and coal to fuel over a million gasoline and diesel vehicles. However, the technology has since been virtually forgotten. This report reviews producer gas and its modern potential. ISBN 0-309-04161-9.

56. **The Diffusion of Biomass Energy Technologies in Developing Countries.** 1984, 120 pp. Examines economic, cultural, and political factors that affect the introduction of biomass-based energy technologies in developing countries. It includes information on the opportunities for these technologies as well as conclusions and recommendations for their application. ISBN 0-309-04253-4.

Technology Options

14. **More Water for Arid Lands: Promising Technologies and Research Opportunities.** 1974, 153 pp. Outlines little-known but promising technologies to supply and conserve water in arid areas. ISBN 0-309-04151-1.

21. **Making Aquatic Weeds Useful: Some Perspectives for Developing Countries.** 1976, 175 pp. Describes ways to exploit aquatic weeds for grazing and by harvesting and processing for use as compost, animal feed, pulp, paper, and fuel. Also describes utilization for sewage and industrial wastewater. ISBN 0-309-04153-X.

34. **Priorities in Biotechnology Research for International Development: Proceedings of a Workshop.** 1982, 261 pp. Report of a workshop organized to examine opportunities for biotechnology research in six areas: 1) vaccines, 2) animal production, 3) monoclonal antibodies, 4) energy, 5) biological nitrogen fixation, and 6) plant cell and tissue culture. ISBN 0-309-04256-9.

61. **Fisheries Technologies for Developing Countries.** 1987, 167 pp. Identifies newer technologies in boat building, fishing gear and methods, coastal mariculture, artificial reefs and fish aggregating devices, and processing and preservation of the catch. The emphasis is on practices suitable for artisanal fisheries. ISBN 0-309-04260-7.

73. **Applications of Biotechnology to Traditional Fermented Foods**. 1992, 207 pp. Microbial fermentations have been used to produce or preserve foods and beverages for thousands of years. New techniques in biotechnology allow better understanding of these transformations so that safer, more nutritious products can be obtained. This report examines new developments in traditional fermented foods. ISBN 0-309-04685-8.

Plants

47. **Amaranth: Modern Prospects for an Ancient Crop.** 1983, 81 pp. Before the time of Cortez, grain amaranths were staple foods of the Aztec and Inca. Today this nutritious food has a bright future. The report discusses vegetable amaranths also. ISBN 0-309-04171-6.

53. **Jojoba: New Crop for Arid Lands.** 1985, 102 pp. In the last 10 years, the domestication of jojoba, a little-known North American desert shrub, has been all but completed. This report describes the plant and its promise to provide a unique vegetable oil and many likely industrial uses. ISBN 0-309-04251-8.

63. **Quality-Protein Maize.** 1988, 130 pp. Identifies the promise of a nutritious new form of the planet's third largest food crop. Includes information on the importance of maize, malnutrition and protein quality, experiences with quality-protein maize (QPM), QPM's potential uses in feed and food, nutritional qualities, genetics, research needs, and limitations. ISBN 0-309-04262-3.

64. **Triticale: A Promising Addition to the World's Cereal Grains.** 1988, 105 pp. Outlines the recent transformation of triticale, a hybrid between wheat and rye, into a food crop with much potential for many marginal lands. The report discusses triticale's history, nutritional quality, breeding, agronomy, food and feed uses, research needs, and limitations. ISBN 0-309-04263-1.

67. **Lost Crops of the Incas.** 1989. 415 pp. The Andes is one of the seven major centers of plant domestication but the world is largely unfamiliar with its native food crops. When the Conquistadores brought the potato to Europe, they ignored the other domesticated Andean crops—fruits, legumes, tubers, and grains that had been cultivated for centuries by the Incas. This book focuses on 30 of the "forgotten" Incan crops that show promise not only for the Andes but for warm-temperate, subtropical, and upland tropical regions in many parts of the world. ISBN 0-309-04264-X.

70. **Saline Agriculture: Salt-Tolerant Plants for Developing Countries**. 1989, 150 pp. The purpose of this report is to create greater awareness of salt-tolerant plants and the special needs they may fill in developing countries. Examples of the production of food, fodder, fuel, and other products are included. Salt-tolerant plants can use land and water unsuitable for conventional crops and can harness saline resources that are generally neglected or considered as impediments to, rather than opportunities for, development. ISBN 0-309-04266-6.

Innovations in Tropical Forestry

35. **Sowing Forests from the Air.** 1981, 64 pp. Describes experiences with establishing forests by sowing tree seed from aircraft. Suggests testing and development of the techniques for possible use where forest destruction now outpaces reforestation. ISBN 0-309-04257-7.

41. **Mangium and Other Fast-Growing Acacias for the Humid Tropics.** 1983, 63 pp. Highlights 10 acacia species that are native to the tropical rain forest of Australasia. That they could become valuable forestry

resources elsewhere is suggested by the exceptional performance of *Acacia mangium* in Malaysia. ISBN 0-309-04165-1.

42. **Calliandra: A Versatile Small Tree for the Humid Tropics.** 1983, 56 pp. This Latin American shrub is being widely planted by villagers and government agencies in Indonesia to provide firewood, prevent erosion, provide honey, and feed livestock. ISBN 0-309-04166-X.

43. **Casuarinas: Nitrogen-Fixing Trees for Adverse Sites.** 1983, 118 pp. These robust, nitrogen-fixing, Australasian trees could become valuable resources for planting on harsh eroding land to provide fuel and other products. Eighteen species for tropical lowlands and highlands, temperate zones, and semiarid regions are highlighted. ISBN 0-309-04167-8.

52. **Leucaena: Promising Forage and Tree Crop for the Tropics.** 1984 (2nd edition), 100 pp. Describes a multipurpose tree crop of potential value for much of the humid lowland tropics. Leucaena is one of the fastest growing and most useful trees for the tropics. ISBN 0-309-04250-X.

71. **Neem: A Tree for Solving Global Problems.** 1992, 149 pp. The neem tree is potentially one of the most valuable of all arid-zone trees. It shows promise for pest control, reforestation, and improving human health. Safe and effective pesticides can be produced from seeds at the village level with simple technology. Neem can grow in arid and nutrient-deficient soils and is a fast-growing source of fuelwood. ISBN 0-309-04686-6.

Managing Tropical Animal Resources

32. **The Water Buffalo: New Prospects for an Underutilized Animal.** 1981, 188 pp. The water buffalo is performing notably well in recent trials in such unexpected places as the United States, Australia, and Brazil. Report discusses the animal's promise, particularly emphasizing its potential for use outside Asia. ISBN 0-309-04159-7.

44. **Butterfly Farming in Papua New Guinea.** 1983, 36 pp. Indigenous butterflies are being reared in Papua New Guinea villages in a formal government program that both provides a cash income in remote rural areas and contributes to the conservation of wildlife and tropical forests. ISBN 0-309-04168-6.

45. **Crocodiles as a Resource for the Tropics.** 1983, 60 pp. In most parts

of the tropics, crocodilian populations are being decimated, but programs in Papua New Guinea and a few other countries demonstrate that, with care, the animals can be raised for profit while protecting the wild populations. ISBN 0-309-04169-4.

46. **Little-Known Asian Animals with a Promising Economic Future.** 1983, 133 pp. Describes banteng, madura, mithan, yak, kouprey, babirusa, javan warty pig, and other obscure but possibly globally useful wild and domesticated animals that are indigenous to Asia. ISBN 0-309-04170-8.

68. **Microlivestock: Little-Known Small Animals with a Promising Economic Future.** 1990, 449 pp. Discusses the promise of small breeds and species of livestock for Third World villages. Identifies more than 40 species, including miniature breeds of cattle, sheep, goats, and pigs; eight types of poultry; rabbits; guinea pigs and other rodents; dwarf deer and antelope; iguanas; and bees. ISBN 0-309-04265-8.

Health

49. **Opportunities for the Control of Dracunculiasis.** 1983, 65 pp. Dracunculiasis is a parasitic disease that temporarily disables many people in remote, rural areas in Africa, India, and the Middle East. Contains the findings and recommendations of distinguished scientists who were brought together to discuss dracunculiasis as an international health problem. ISBN 0-309-04172-4.

55. **Manpower Needs and Career Opportunities in the Field Aspects of Vector Biology.** 1983, 53 pp. Recommends ways to develop and train the manpower necessary to ensure that experts will be available in the future to understand the complex ecological relationships of vectors with human hosts and pathogens that cause such diseases as malaria, dengue fever, filariasis, and schistosomiasis. ISBN 0-309-04252-6.

60. **U.S. Capacity to Address Tropical Infectious Diseases.** 1987, 225 pp. Addresses U.S. manpower and institutional capabilities in both the public and private sectors to address tropical infectious disease problems. ISBN 0-309-04259-3.

Resource Management

50. **Environmental Change in the West African Sahel.** 1984, 96 pp. Identifies measures to help restore critical ecological processes and

thereby increase sustainable production in dryland farming, irrigated agriculture, forestry and fuelwood, and animal husbandry. Provides baseline information for the formulation of environmentally sound projects. ISBN 0-309-04173-2.

51. **Agroforestry in the West African Sahel.** 1984, 86 pp. Provides development planners with information regarding traditional agroforestry systems—their relevance to the modern Sahel, their design, social and institutional considerations, problems encountered in the practice of agroforestry, and criteria for the selection of appropriate plant species to be used. ISBN 0-309-04174-0.

72. **Conserving Biodiversity.** A Research Agenda for Development Agencies. 1992, 127 pp. Reviews the threat of loss of biodiversity and its context within the development process and suggests an agenda for development agencies. ISBN 0-309-04683-1.

Forthcoming Books from BOSTID

Vetiver Grass for Soil and Water Conservation. (1992) This study will evaluate the potential of vetiver, a little-known grass that seems to offer a practical solution for controlling soil loss. Hedges of this deeply rooted grass catch and hold back sediments. The stiff foliage acts as a filter that also slows runoff and keeps moisture on site, allowing crops to thrive when neighboring ones are desiccated. In numerous tropical locations, vetiver hedges have restrained erodible soils for decades and the grass—which is pantropical—has shown little evidence of weediness.

BOSTID Publication Distributors

U.S.:

AGRIBOOKSTORE
1611 N. Kent Street
Arlington, VA 22209

AGACCESS
PO Box 2008
Davis, CA 95617

Europe:

I.T. PUBLICATIONS
103-105 Southhampton Row
London WC1B 4HH
Great Britain

S. Toeche-Mittler
TRIOPS Department
Hindenburgstr. 33
6100 Darmstadt
Germany

T.O.O.L. PUBLICATIONS
Sarphatistraat 650
1018 AV Amsterdam
Netherlands

Asia:

ASIAN INSTITUTE OF TECHNOLOGY
Library & Regional
Documentation Center
PO Box 2754
Bangkok 10501
Thailand

NATIONAL BOOKSTORE
Sales Manager
PO Box 1934
Manila
Philippines

UNIVERSITI OF MALAYA COOP. BOOKSHOP LTD.
Universiti of Malaya
Main Library Building
59200 Kuala Lumpur
Malaysia

RESEARCHCO PERIODICALS
1865 Street No. 139
Tri Nagar
Delhi 110 035
India

CHINA NATL. PUBLICATIONS
IMPORT & EXPORT CORP.
PO Box 88F
Beijing
China

South America:

ENLACE LTDA.
Carrera 6a. No. 51-21
Bogota, D.E.
Colombia

Africa:

TAECON
c/o Agricultural Engineering Dept
P.O. Box 170 U S T
Kumasi
Ghana

Australasia:

TREE CROPS CENTRE
P.O. Box 27
Subiaco, WA 6008
Australia

For More Information

To receive more information about BOSTID reports and programs, please fill in the attached coupon and mail it to:

Board on Science and Technology for International Development
Publications and Information Services (FO-2060Z)
Office of International Affairs
National Research Council
2101 Constitution Avenue, N.W.
Washington, D.C. 20418 USA

Your comments about the value of these reports are also welcome.

- -

Name ______________________________
Title ______________________________
Institution ______________________________
Street Address ______________________________
____
City ______________________________
Country ______________ Postal Code ______________

73

- -

Name ______________________________
Title ______________________________
Institution ______________________________
Street Address ______________________________
____
City ______________________________
Country ______________ Postal Code ______________

73

- -

Name ______________________________
Title ______________________________
Institution ______________________________
Street Address ______________________________
____
City ______________________________
Country ______________ Postal Code ______________

73